The evolution of air breathing in vertebrates

The evolution of air breathing in vertebrates

DAVID J. RANDALL

Zoology Department
University of British Columbia

WARREN W. BURGGREN

Zoology Department
University of Massachusetts

ANTHONY P. FARRELL

Cardiovascular Research Laboratory
University of Southern California Medical Center

M. STEPHEN HASWELL

Department of Zoology and Entomology
Colorado State University

CAMBRIDGE UNIVERSITY PRESS

Cambridge

London New York New Rochelle

Melbourne Sydney

Published by the Press Syndicate of the University of Cambridge
The Pitt Building, Trumpington Street, Cambridge CB2 1RP
32 East 57th Street, New York, NY 10022, USA
296 Beaconsfield Parade, Middle Park, Melbourne 3206, Australia

First published 1981

Printed in the United States of America
Typeset by Bi-Comp Inc., York, Pa.
Printed and bound by The Murray Printing Company, Westford, Mass.

Library of Congress Cataloging in Publication Data

Main entry under title:

The Evolution of air breathing in vertebrates.

Bibliography: p.

Includes index.

1. Respiration.
2. Lungs – Evolution.
3. Gills – Evolution.
4. Vertebrates – Evolution.

I. Randall, David J., 1938–
QP121.E96 596′.01′2 80–462
ISBN 0 521 22259 1

Contents

Preface

Many physiological studies are interpreted in anthropocentric terms, a tendency reinforced by our more detailed understanding of mammals. This is unfortunate because mammals, with their large size, high metabolic rate, and complete dependence on air breathing are very specialized animals that have invaded only a relatively small section of the total environment. Nevertheless, because of anthropocentric tendencies, studies of the evolution of air breathing in vertebrates have often attempted to trace the process backward from mammals rather than forward in time from aquatic ancestors.

The problem of describing the evolution of air-breathing vertebrates starting with the ancestral aquatic forms is exacerbated by our complete lack of physiological knowledge of these fish. These ancient fish, however, did give rise to modern fish, as well as terrestrial vertebrates, and we can achieve some understanding of these ancestral stocks by studying their aquatic as well as terrestrial descendants.

In this book we decided to try to describe the aquatic ancestral form first and then attempt to delineate the changes that must have taken place during the evolution of air breathing. This was as much a consequence of history as philosophy, as my initial experimental interest was in completely aquatic vertebrates. The results of these earlier studies raised some questions about the types of changes that must have occurred as vertebrates invaded land. We initiated a series of studies on bimodally breathing fish and amphibians in order to answer these questions. In the process three of us were fortunate enough to visit the Amazon river, a setting that I think gives perspective to the problem, and which provided an experience I found most instructive. Our experiments on bimodally breathing vertebrates are still in progress, and this book represents not so much a review of the literature as our present view of the nature of the process of evolution of air breathing in vertebrates.

David Randall

Vancouver, Canada

1 *Introduction: air breathing in vertebrates*

"Biological innovations tend to appear soon after environmental conditions become favorable to them" (Cloud 1974). Since the Paleozoic era conditions have existed favoring the evolution of air breathing in vertebrates, and many species have evolved over this span of between 500 and 600 million years.

The central theme of this book is an analysis of the changes in physiological processes that occurred as air-breathing, terrestrial vertebrates evolved from aquatic forms. Air breathing has evolved many times and was not a single event. Were the changes similar in all cases, and can we generalize, in a functional sense, about the evolution of all air-breathing vertebrates? Although there is no direct physiological evidence contained in the paleontological record, and we can only study species alive today, we believe a number of generalizations can be made. The basis for our belief is as follows. All air-breathing vertebrates have evolved from the same subclass of aquatic vertebrates, the Osteichthyes, the characteristics of which must set limits to the nature of the process. In addition, a major selective force in the evolution of air-breathing forms appears to have been aquatic hypoxia (see Packard 1974). Finally, the range of available habitats in the aquatic environment has remained the same, but with large changes in spatial distribution, over the time span of vertebrate evolution (Holland 1975; Holser 1977). Thus, all air-breathing vertebrates have evolved from the same group of aquatic vertebrates, existing in a similar range of environments, as the result of the operation of similar selective forces. It seems unlikely, therefore, that the nature of the process would be different in each case. The evolution of air-breathing vertebrates appears to be a classic case of parallel evolution.

The purely aquatic Osteichthyes may have changed markedly in their general organization since the Devonian period. There are clear changes in skeletal structure in successive groups, but how much has their physiological and biochemical organization changed? No clear-cut answer can be given because no direct information is available in the paleontological record. Thus it is difficult to determine to what extent the gills of modern fish are similar to those of their Devonian ancestors. If the same characteristics are found in two groups of animals with a common ancestor, then either convergent evolution has occurred, or those characteristics also belonged to the ancestor. For example, water and sodium turnover rates, which are related to the characteristics of the gill epithelium, are similar in Chondrostei (Potts and Rudy 1972), Dipnoi (Oduleye 1977), and Teleostei (Maetz 1973), and both elasmobranchs and teleosts have Na^+/H^+ or Na^+/NH_4^+ and HCO_3^-/Cl^- exchange processes in their gills (Payan and Maetz 1973; Maetz 1973; Bornancin et al. 1977). Elasmobranchs and teleosts have been isolated since the Devonian period, and it seems probable that the general functional organization of the gills in ion and water regulation is fundamental to both groups and was present in the antecedents of all air-breathing vertebrates. The teleost gill is the most thoroughly understood of any aquatic gas-exchange organ. Although there are clear gross morphological differences and minor quantitative variations in functional properties, it appears reasonable to us to assume that the functional organization of the teleost gill can be taken as qualitatively representative of the mode of function of the gills of all aquatic gnathostomes since the Devonian period.

Aquatic ancestors of air-breathing vertebrates

The gnathostome vertebrates separated into three distinct groups, the Placodermi, Chondrichthyes, and Teleostomes (Figure 1.1), at some time in the Silurian period. The placoderms and the two divisions of the Chondrichthyes, the elasmobranchs and holocephalans, do not appear to have gas bladders or to have given rise to any air-breathing forms (Moy-Thomas 1971). The placoderm *Bothriolepis* is a possible exception.

The Osteichthyes or bony fish includes the Crossopterygii, Dipnoi, and Actinopterygii, and all members characteristically have a gas bladder of some description during at least some period of their life. The crossopterygians, like lungfish and tetrapods, have a lung with a ventral pharyngeal opening, but the actinopterygians, with the exception of polypterids, have a gas bladder with a duct opening dorsally from the gut.

It is probably not simply coincidence that the aquatic ancestors of air-breathing vertebrates had both air bladders and bony skeletons. The high density of bone can be offset by buoyant air bladders. The question of whether the ubiquitous air bladder of the crossopterygians functioned initially in gas transfer or buoyancy control is difficult to answer, and the processes are clearly not mutually exclusive. An air bladder used for gas transfer will also make the fish more buoyant, and skeletal density can be increased to bring the fish closer to neutral buoyancy. A gas bladder used for buoyancy control can be a source of oxygen during periods of oxygen lack. What is clear is that aquatic ancestors of air-breathing vertebrates had a gas bladder and bony skeletons.

All air-breathing vertebrates have evolved from bony fish, and it is interesting to speculate why the elasmobranchs have not evolved any air-breathing forms. We think the answer is probably twofold. First, this group has occupied either a benthic or nektonic niche within the aquatic environment. The benthic forms are distant from the air–water interface, and the open ocean is seldom hypoxic. Second, in elasmobranchs there has been a tendency to reduce density by including a less bony skeleton and increasing lipid content rather than including any gas floats such as bladders. Thus elasmobranchs exist in well-aerated water, lack gas bladders, and have a form and function such that they seldom cross the air–water interface.

Most of the surviving genera of largely extinct groups of bony fish are air-breathing (Figure 1.2), and in all these phylogenetic relic species gas bladders or lungs are the organs for air breathing. Surviving genera from groups of air-breathing fish other than teleosts tend to be obligatory air breathers, at least during some part of their life cycle. Many of these fish can withstand prolonged periods of drought. Some lungfish aestivate in cocoons within mud burrows and can survive many months of dehydration, as can the bowfin, *Amia*. Living specimens of the latter have been dug up in fields that at one time had been flooded but then were dry for several years (Dence 1933). This combination of bimodal breathing with a well-developed lung or gas bladder plus the ability to survive dehydration has enabled many species of chondrosteans, nonteleost neopterygians, and Dipnoi to survive where aquatic relatives have been largely eliminated by competition with successive waves of water-breathing bony fish. Selective advantage for strictly water-breathing fish is largely related to biomechanical efficiency in locomotion, and this has determined the success of each new wave of aquatic vertebrates. These forces appear to be less critical in air-breathing fish, which have tended to be away from the action both biologically and geographically. They are animals specialized for survival in a particular environment, the air–water interface, which imposes quite different demands.

Environmental history

Life evolved in a chemically reducing atmosphere of methane, ammonia, and hydrogen (Cloud 1974; Tappan 1974; Schidlowski 1975). The production of oxygen in the environment was the result of the evolution of organisms, similar to blue-green algae, capable of photosynthetic activity. After the onset of photosynthetic activity, however, oxygen levels remained low for a considerable period of time because any oxygen produced was rapidly utilized in oxidation reactions in the substrates. In the original anoxic world the waters contained high levels of soluble ferrous iron, and, in the reducing atmosphere of methane, iron did not rust. Thus the initial production of oxygen was probably utilized in the oxidation of ferrous to ferric iron, which then precipitated and produced the Banded Iron Formations found in the geological record. By the Precambrian period, however, oxygen levels in the atmosphere were probably around 3% of present levels (Cloud 1974). This rise in atmospheric oxygen heralded the appearance of the

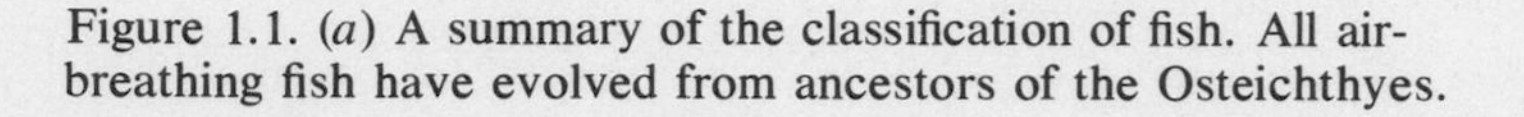
Figure 1.1. (*a*) A summary of the classification of fish. All air-breathing fish have evolved from ancestors of the Osteichthyes.

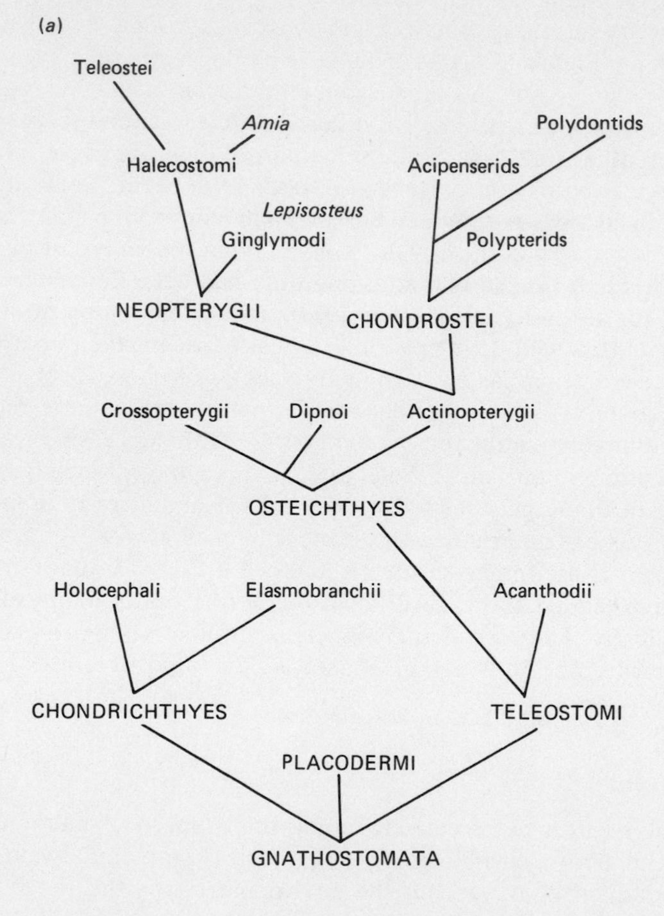

Metazoa and contributed to the rapid series of evolutionary events that had occurred by the Cambrian period.

The first fish appeared in the Cambrian period and radiated extensively in the Silurian and Devonian periods. The oxygen content of the atmosphere was probably high, and may have been at more than present levels in the late Cambrian through to the Devonian period (Figure 1.3).

There were probably five continents on Earth in the Cambrian period (Palmer 1974), consisting of (1) present-day North America, (2) Northern Europe, (3) Siberia, (4) Mongolia plus China, and (5) Gondwanaland. The last was an extensive southern continent consisting of present-day South America, Australia, Africa, and Antarctica. Conti-

Figure 1.1. (*b*) Romer's (1972) view of the Osteichthyes. Forms in which lungs are known to exist are underlined. Hatched areas indicate where lungs are presumed to be present and include ancestral forms of the bony fish stock.

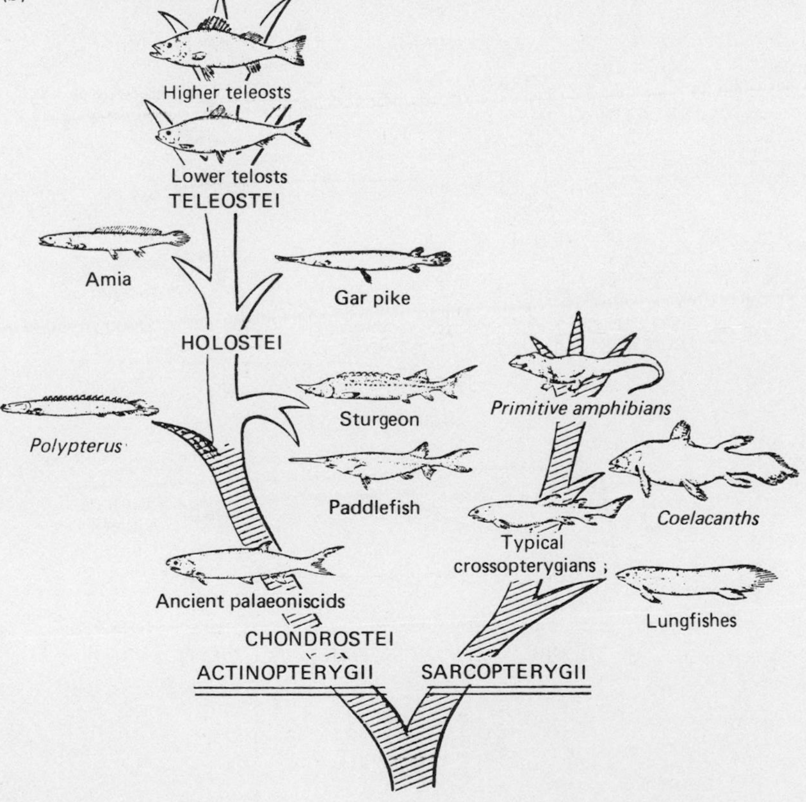

nental drift over the next 200–300 million years rearranged these land masses so that, by the Permian period, the single continent Pangaea existed. Pangaea then broke up into our present continental pattern.

The Earth's climate was probably warm and equable from the Cambrian to the Silurian period, and there was essentially continual summer with large equatorial regions and little ice or snow at the poles. Days and nights were shorter in the Cambrian period, each year consisting of

Figure 1.2. The time sequence of the evolution of various terrestrial and aquatic vertebrate groups. The thickened regions indicate large numbers of species of that group at that time.

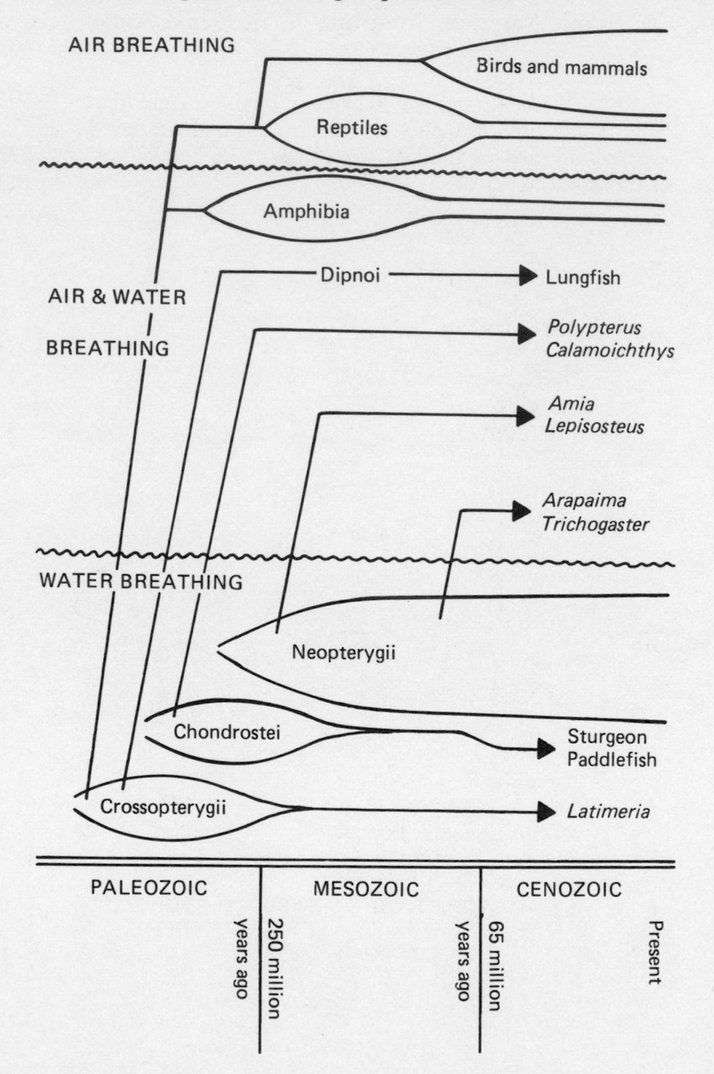

425 days. This was reduced to 400 days per year by the Devonian period as the Earth's speed of rotation decreased. The reduced day length presumably influenced environmental conditions, and circadian oscillations in physical parameters, temperature for instance, were probably reduced. The halcyon conditions in the extensive warm shallow seas of the late Silurian period supported the emergence of large numbers of invertebrate and vertebrate genera. Some plants and a few arthropods had become established on land.

During the Devonian period the northern continents remained warm and equable, but primitive gymnosperms, like *Callixylan* in forests around present-day New York, show annual rings indicating seasonal growth. Gondwanaland, in southern regions, may have been glaciated. In the late Devonian, extending through the Carboniferous and Permian periods the climate cooled, more land was exposed, and land masses became more arid. Oxygen levels in the atmosphere and water may also have dropped precipitously (Figure 1.3). This combination of catastrophies resulted in the elimination of 90% of the fish species in the late Devonian period and a further loss of 50% of fish families by the end of the Permian period. Thus, the Paleozoic era was a period of stability and growth followed by what Tappan (1974) has referred to as the "Mid-Paleozoic crisis," finally ending in the cool, arid conditions of the Permian period.

Temporal variations in the extent of glaciation have undoubtedly altered water levels and salinity, but they must have been small compared with spatial variations between water bodies at any one time. For instance, the range of salinities present today in natural bodies of water

Figure 1.3. Possible fluctuations in atmospheric oxygen levels over the period of vertebrate evolution. The reduction in oxygen in the Devonian period was associated with the extinction of many species and the appearance of many air-breathing vertebrates; (PAL) present atmospheric level (from Tappan 1974).

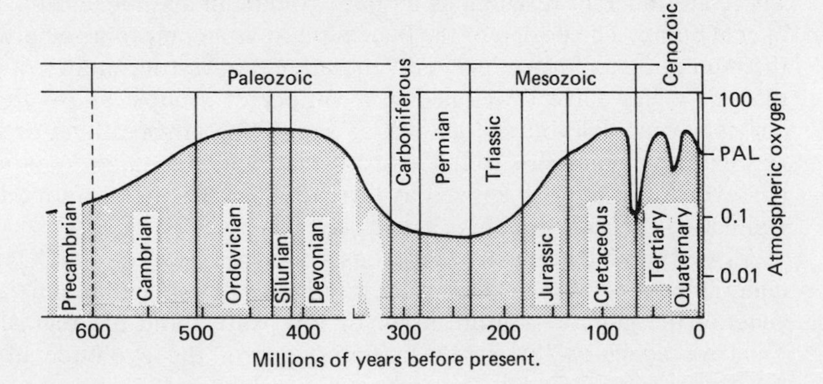

varies from nearly distilled water in some glacial lakes to the hypersaline conditions of the Dead Sea or the Salt Lake in Utah. By comparison, temporal changes in seawater composition since the Precambrian period have been minor (Holland 1975).

The evolution of air-breathing forms

Air-breathing teleosts inhabit both freshwater and intertidal environments. Although most remnant fish species are freshwater, and the crossopterygians are generally regarded as being a freshwater group since the Mid-Devonian, there is no reason to regard salinity as an important factor limiting the evolution of air-breathing vertebrates (Thomson 1969; Packard 1974). Aquatic hypoxia is probably the more important selective force in the evolution of air-breathing vertebrates, and it is simply more characteristic of freshwaters. Regions of local aquatic hypoxia, due to metabolic oxygen consumption and organic decay, have probably existed since the Cambrian period. Lungfish, which are able to survive hypoxic water and seasonal drought, have changed little since the Carboniferous period. This fact indicates the regional persistence of aquatic hypoxia since the Paleozoic era. There have probably been large oscillations in atmospheric oxygen levels over more recent times (Figure 1.3), and these changes will have been duplicated in the water. Under these conditions there is clearly no refuge from hypoxia. Even so, the O_2 content of the air is still higher than that of water, and air breathing is of selective advantage.

We envisage that the following sequence of events occurred in the evolution of air breathing in vertebrates. In conditions of aquatic hypoxia there is a clear selective advantage to those animals with some means of utilizing oxygen from the atmosphere. The simplest means for a fish to ventilate an air-breathing organ is to utilize the buccal pump, which is used for gill ventilation. Thus the first step was probably the evolution of an air-breathing organ ventilated by a buccal pump. There are inherent restrictions on tidal volume in an animal utilizing a buccal pump. The design of the buccal pump is a compromise between the requirements for feeding and ventilation. Aspiratory modes of ventilation, which have developed in a variety of groups, allow greater control over tidal volume, as well as separating air-breathing organs' ventilatory mechanisms from those associated with feeding. Aspiration has evolved in aquatic as well as terrestrial vertebrates as a mode of ventilating their air-breathing organ (see Chapter 4).

All air-breathing organs are used as an oxygen source, but CO_2 normally is not excreted in large quantity into air in fish and amphibia. In general, the gills or skin are used for ion, water, and pH regulation. Thus we conclude that in the initial stages of the evolution of air-breathing, bimodal breathing is required (Figure 1.4). The air-breathing organ is used for oxygen uptake, and the aquatic respiratory surface is used for CO_2 excretion and pH regulation. The relocation of ion and H^+ regulation from the gills to the skin was a subsequent step in the evolution of air breathing, and its elucidation requires an understanding of the coupling of ion and H^+ regulation in the gills of aquatic vertebrates (Chapter 2).

In reptiles, birds, and mammals there is a marked change from the aquatic pattern of CO_2 excretion. Here, the lungs play an important role in acid–base regulation by controlling CO_2 levels in the blood via ventilation. This enables the animal to become less dependent on water for ion and H^+ regulation and removes the requirement for bimodal breathing. The requirements for gill ventilation and/or wetting the skin disappeared, and the removal of the cooling effect of water on the surface, owing to its high specific and evaporative heat, reduced surface

Figure 1.4. A schematic arrangement of vertebrate groups indicating functional steps in the evolution of air-breathing forms. The terms *buccal* and *aspiratory* refer to the mode of ventilation of the respiratory structure.

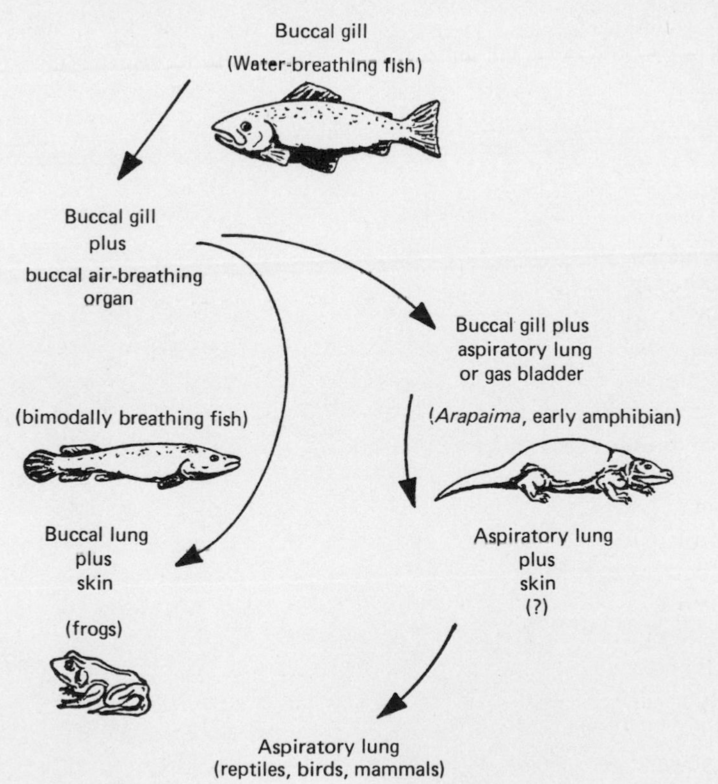

cooling. Heat loss could be controlled and body temperature could be regulated at a level above ambient. The evolutionary changes in the vertebrate circulatory system (Chapter 3) reflect this sequence of events as aquatic forms evolved from unimodal water breathers into bimodally breathing amphibious forms to unimodal air-breathing animals (Figure 1.4).

Clearly all extant species are "successful," and birds and mammals are not the "ultimate" in vertebrate evolution. The lungfish and various remnant species of the Actinopterygii, retaining a bimodal breathing pattern, have evolved attributes allowing them to occupy and hold a niche successfully. Although unimodal breathing permits radiation in either water or air, the animal is restricted to one phase only. Bimodal breathing has allowed a number of vertebrates to hold onto a niche successfully at the interface of air and water against the most sophisticated of unimodally breathing vertebrates. Analysis of their pattern of gas transfer gives us some insight into the nature of the process in the transition of vertebrates from water to land (Chapter 2).

2 *Gas transfer: the transition from water to air breathing*

Air versus water as a respiratory medium

Air-saturated water contains about one-thirtieth of the oxygen found in an equal volume of air; even so, most vertebrates, including water-breathing fish, can maintain oxygen delivery to the tissues in the face of changes in both oxygen supply and demand. There is an ultimate level of hypoxia, however, at which an animal is unable to maintain oxygen delivery, and aerobic metabolism falls. The exact level at which this occurs varies widely with the species. When environmental oxygen levels become limiting, behavioral, physiological, and/or biochemical mechanisms are invoked which enable some animals to survive hypoxic conditions. In general, the behavioral mechanisms are to escape from the hypoxic conditions. Many fish avoid hypoxic waters by moving to skim the surface waters where the oxygen content is usually higher. Some fish species breathe air and tap the generally stable and large oxygen supply in the atmosphere. Some of these air-breathing fish, particularly those that inhabit waters that are nearly always hypoxic, are obligate air breathers. For example, the pirarucu, *Arapaima gigas,* relies on air breathing for most of its oxygen supply, and will die if not allowed access to the surface (Stevens and Holeton 1978).

Carbon dioxide levels in lakes and oceans vary considerably, unlike atmospheric levels, which remain very low and stable. These oscillations in CO_2 levels create problems for aquatic animals, affecting CO_2 excretion and therefore hydrogen ion regulation. Aquatic vertebrates, however, have evolved mechanisms for maintaining $[H^+]$ in the face of varying environmental CO_2 levels without resorting to air breathing and the utilization of the atmosphere as a sink for carbon dioxide (see section on hydrogen ion regulation). It appears, therefore, that air breathing in vertebrates evolved in response to aquatic hypoxia, and not to aquatic hypercapnia. In most fish, air breathing, or more particularly air exposure, creates rather than solves problems of CO_2 excretion and H^+ regulation. In fact, most air-breathing fish and many amphibians must remain in water or keep a wet body surface in order to control CO_2 excretion and regulate $[H^+]$ in the body. Changes in the pattern of CO_2 excretion are central to understanding the transition from aquatic to aerial breathing.

Gas transfer systems

Aquatic vertebrates have gills; air-breathing vertebrates typically have lungs. The evolution of air-breathing vertebrates from aquatic ancestors clearly involves the evolution of lungs and the disappearance of gills. In order to analyze this process in detail we first must consider the characteristics of gas transfer in both gills and lungs.

The mammalian lung

The dimensions of the alveolar membrane and the pattern of air and blood flow in the lung are such that both O_2 and CO_2 in blood and alveolar gas rapidly attain equilibrium. The oxygenation of hemoglobin is very rapid, and the formation of CO_2 from bicarbonate is catalyzed by carbonic anhydrase, such that any reactions of gases within the blood, and diffusion of gas between air and blood, are complete in approximately 250–500 msec (West 1974). The blood is resident in the lung for longer periods, probably residing for at least 250 msec even during maximum exercise (Figure 2.1). Thus, conditions in the lung are

Figure 2.1. Oxygen time courses in the pulmonary capillary when diffusion is normal and abnormal (e.g., because of thickening of the alveolar membrane). Note that exercise reduces the time available for oxygenation, but that it is essentially complete in the normal lung (from West 1974).

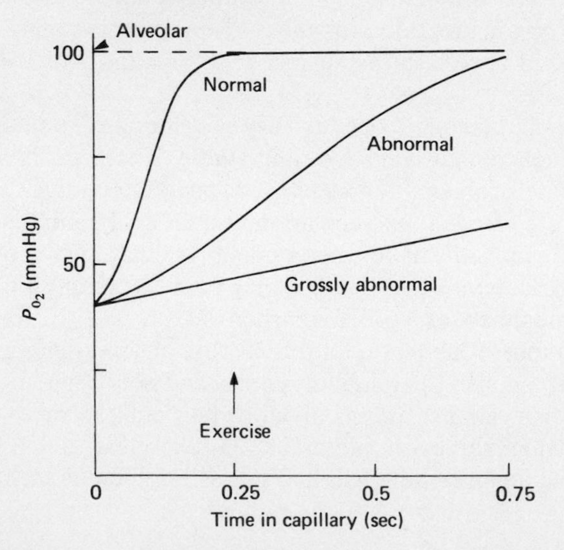

such that neither diffusion nor any reaction velocities limit the transfer of oxygen or carbon dioxide; that is, gas transfer is not *diffusion-limited* (in general, limitations due to diffusion and those due to slow reaction velocities are difficult to separate, so they are lumped together and considered as diffusion limitations). Mammals can increase the rate of gas transfer by simply increasing blood perfusion and lung ventilation. Changes in ventilation normally follow and equal lung perfusion, which is proportional to oxygen and carbon dioxide transfer. Thus the mammalian lung is often considered to be *perfusion-limited.* This applies to both oxygen uptake and CO_2 removal, and increases in lung perfusion and ventilation will lead to an increase in both O_2 and CO_2 transfer (but in opposite directions).

Equal volumes of air and mammalian blood at ambient P_{O_2} contain about the same amount of oxygen, and lung ventilation and pulmonary blood flow in mammals are equal; that is, the ventilation : perfusion ratio is about one (Figure 2.2). There is a large residual volume in the lungs of mammals, and a small volume of air is mixed with this residual volume at each breath; hence the alveolar P_{O_2} is around 100 mmHg,

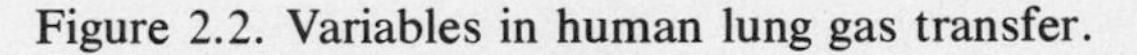

Figure 2.2. Variables in human lung gas transfer.

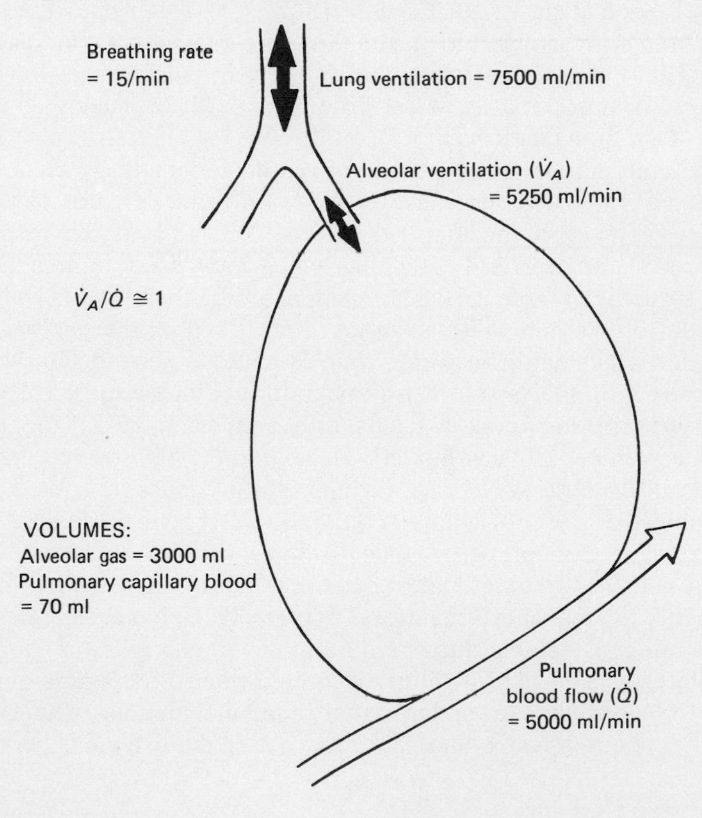

and P_{CO_2} is about 40 mmHg, oscillating by a few millimeters of mercury with each breath. An increase in ventilation without change in oxygen utilization or CO_2 production will reduce alveolar gas P_{CO_2} and therefore arterial blood P_{CO_2} and raise P_{O_2}. The rise in blood P_{O_2} will increase the concentration of oxygen in solution, but, as hemoglobin in arterial blood is usually fully saturated with oxygen, there will be only a small increase in blood oxygen content. The reduction in P_{CO_2} will have a marked effect on blood pH. Air-breathing mammals can in this way adjust blood P_{CO_2} levels by changing lung ventilation in proportion to CO_2 production and thereby regulate pH. This is a rapid and powerful mechanism for regulating body [H^+] and is not available to water-breathing vertebrates.

The fish gill

Water contains much less dissolved oxygen than an equal volume of air; it is some 1000 times more dense and viscous, and gases diffuse 10^5 times more slowly in water than in air. Gills consist of a sievelike structure perfused with blood and ventilated with a unidirectional flow of water (Figure 2.3). In general, a thin sheet of water flows over a respiratory epithelium in a countercurrent arrangement to a thin sheet flow of blood (Figure 2.4). Because the water contains so little oxygen compared with arterial blood, the flow of water must be much greater than that of blood. In aquatic animals the ventilation : perfusion ratio is often 10 or more; that is, water flow is an order of magnitude greater than blood flow (see Chapter 3).

The body temperatures of water-breathing animals are close to ambient, and at these generally reduced temperatures, compared with mammals, oxygen consumption is less. The area of the respiratory surface is also less, but is similar when expressed as area per unit oxygen uptake (Table 2.1). The residence time for blood in the gills is difficult to determine with accuracy, but, based on morphological and cardiac output data, residence time is about a second (Haswell and Randall 1978), the same order of magnitude as in the mammalian lung.

Oxygen transfer across fish gills, unlike the mammalian lung, is diffusion- as well as perfusion-limited. Fisher et al. (1969) showed that in the bullhead, *Ictalurus nebulosus,* during rest $\dot{M}_{CO}$ (mass transfer of carbon monoxide) was proportional to P_{CO} in the water, and calculated that O_2 uptake, like CO uptake, was also diffusion-limited. Certainly arterial blood and inspired water never achieve equilibrium across fish gills (Randall 1970a), and Scheid and Piiper (1976) have calculated that there are considerable diffusion limitations in fish gills.

Why then is the fish gill diffusion-limited when it has a ratio of surface area to oxygen uptake similar to that found in mammals, where oxygen transfer is not diffusion-limited? First, not all the fish gill is perfused at rest, with only about 60% of the lamellae perfused in a resting trout (Booth 1978). Second, the diffusion distance between the medium and blood is larger in fish than in mammals (Table 2.1). Finally, although the dead-space volume is much reduced in gills compared with lungs, the boundary layer of water next to the respiratory surface may constitute an important added diffusion limitation (Hills 1972).

The gill epithelium is generally much thicker than that of the lung and is involved in ion and water as well as gas transfer. The gill is covered by metabolically active flat epithelial cells with high levels of carbonic anhydrase activity and a capacity for both cation (Na^+/H^+ or NH_4^+) and anion (HCO_3^-/Cl^-) exchange. Six to ten percent of the cells are specialized, mitochondria-rich, chloride cells, which are involved in ion and perhaps water movements across the epithelium. The gills also contain mucous cells, which supply a liberal coating of mucus over the gill surface. There are no surfactants in the gills; however, some fish, notably air-breathing fish and those like the gourami that construct bubble nests, produce a surpellic substance to reduce surface tension at the air–liquid interface (Phleger and Saunders 1978).

Fish can increase oxygen uptake by a factor of 8 to 10 during exercise (Jones and Randall 1978) and can maintain oxygen uptake in the face of aquatic hypoxia (Holeton and Randall 1967). These changes are associated not only with changes in gill water and blood flow but also with changes in the conditions for diffusion in the gills. Fisher et al. (1969), for example, showed that the carbon monoxide–diffusing capacity of bullhead gills increased during hypoxia.

The gill diffusing capacity can be changed in a number of ways. The

Table 2.1. *Conditions for O_2 transfer, fish: mammal*

	Fish	Mammal
	Trout 10–15°C	Man 37°C
Body weight (kg)	0.2	55
Respiratory surface area (m^2)	0.06	63
Respiratory surface area/g body weight	2.97 cm^2/g (1.78 cm^2/g)[a]	11.5 cm^2/g
M_{O_2} at rest (ml/g/hr)	0.04	0.23
Surface area per unit $\dot{M}_{O_2}$	74 (45)[a]	50
Diffusion distance medium/blood	5 μm	1 μm
Residence time for blood at respiratory surface	~1 sec	~1 sec

[a] Function area is only 60% of total area at rest (see text).

secondary lamellae are arranged like the rungs of a ladder on both sides of each gill filament (Figure 2.3). The number of lamellae perfused and, therefore, the functional area available for gas transfer increase during hypoxia (Booth 1978) and probably also during exercise. This increase would decrease the resistance to diffusion across the gills. Furthermore, a portion of each secondary lamella is buried beneath the surface of the filament, and so there are large diffusion distances between the water and blood in these sunken basal channels (see Figure 2.3e). Farrell (1979) has shown that blood flow through the secondary lamellae can be described by sheet flow dynamics (Sobin et al. 1972). Thus the

Figure 2.3. The teleost gill. (*a*) The position of the gills in the fish. (*b*) Schematic of a portion of the gills.

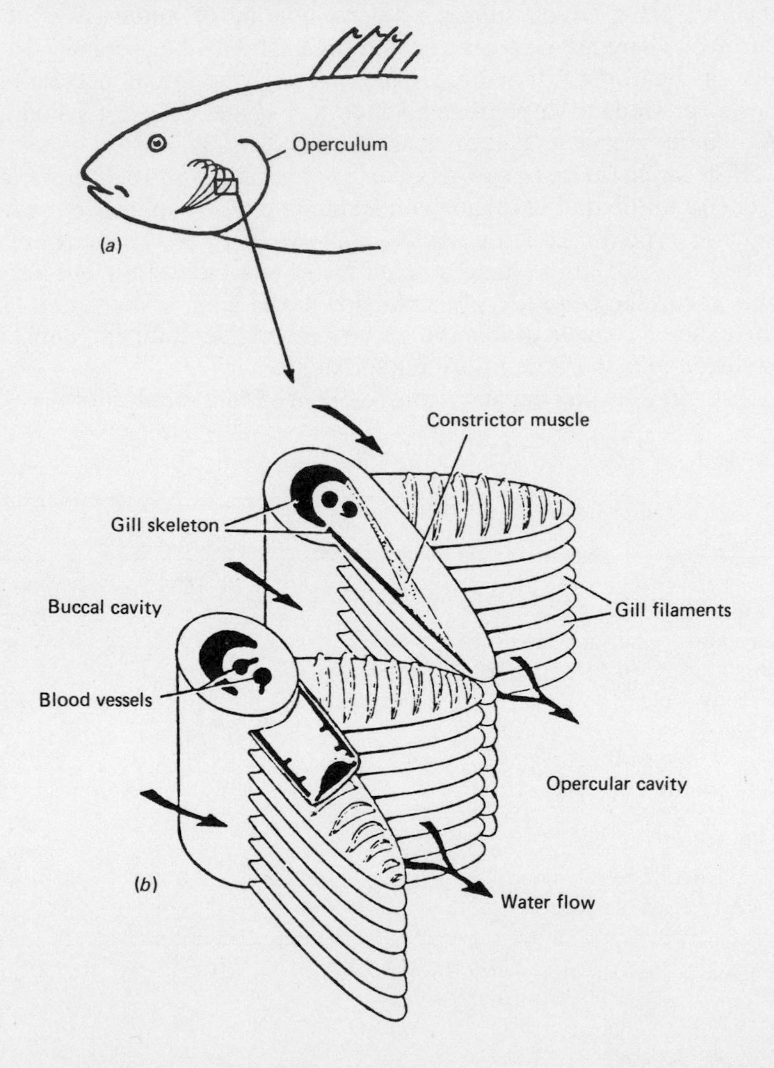

lamellar blood channels, like the lung blood vessels, are very compliant. Small changes in pressure cause a marked redistribution of blood within the secondary lamellae. A rise in pressure increases lamellar volume, and a larger portion of blood flow is directed through sec-

Figure 2.3. (*c*) A plastic cast of a portion of a gill filament. (*d*) A plastic cast of a gill lamella. (*e*) Section through several lamellae. (*f*) Segment of a lamella (high-power section of *d*). Sections and plastic casts are from the lingcod, *Ophiodon elongatus*. The lamellar sections shown in (*e*) were filled with silicone elastomer according to the method of Sobin et al. (1972). Parts (*a*) and (*b*) are after Hughes (1964).

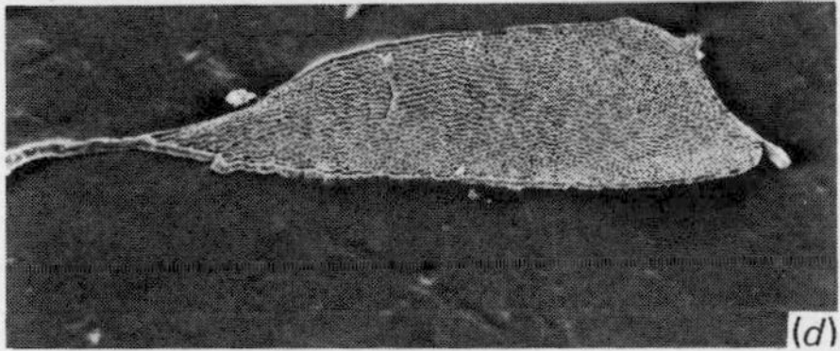

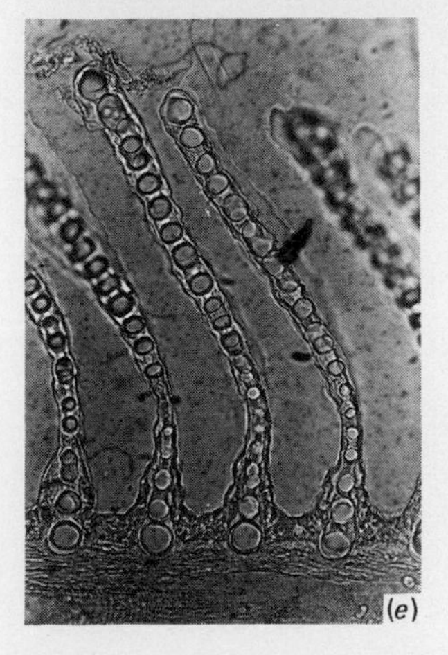

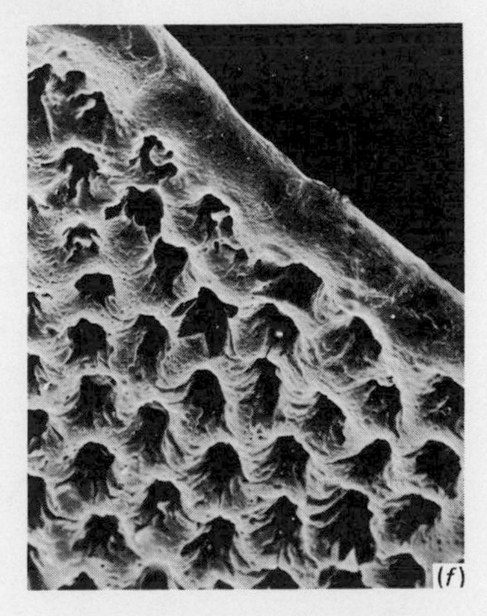

tions of lamellae exposed to water flow. Also vascular changes may cause reductions in the lymphatic space with the lamellae (Figure 2.3). All of these changes result in a reduction in the diffusion distance between blood and water and an increase in available area for gas transfer, and thus increase the diffusing capacity of the gills.

Thus the number of lamellae perfused and their individual pattern of perfusion are controlled such that the gill diffusing capacity as well as ventilation and perfusion can be regulated to meet the oxygen requirements of the tissues. The diffusing capacity of the gills is correlated to oxygen transfer requirements such that the gill diffusing capacity is reduced during low levels of activity in order to limit ion and water transfer (Randall et al. 1972; Wood and Randall 1973).

Oxygen and carbon dioxide transport in blood

There are marked changes in the circulation, particularly in the heart and arterial system, associated with the evolution of air breathing (see Chapter 3), but there are only quantitative differences in the transport

Figure 2.4. Variables in gill gas transfer (trout, 200 g body weight, 8°C).

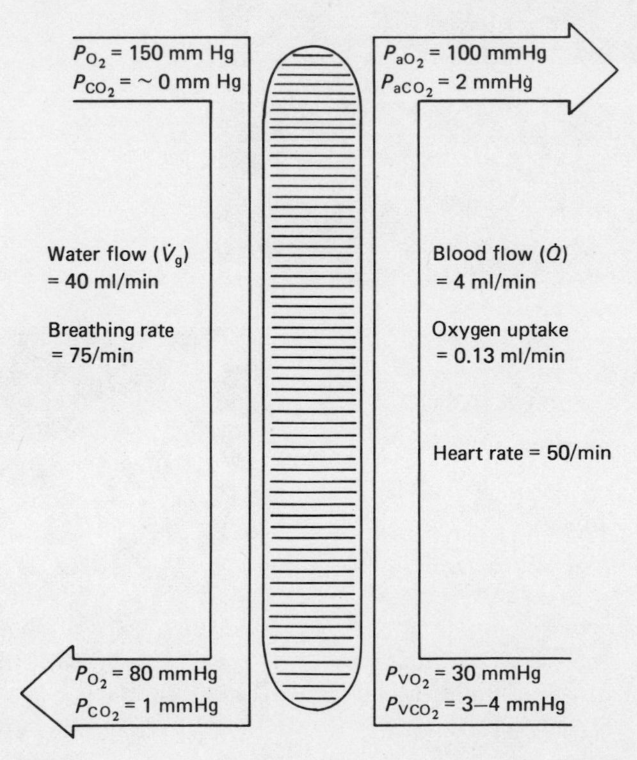

of oxygen and carbon dioxide in the blood between water- and air-breathing vertebrates. In both groups almost all the oxygen transported in the blood is bound to hemoglobin (a notable exception are ice fish, which lack hemoglobin), and plasma bicarbonate is the major pool of CO_2.

$$CO_2 + H_2O \rightleftharpoons H_2CO_3 \rightleftharpoons H^+ + HCO_3^- \qquad (1)$$

Amino groups are found in blood proteins like hemoglobin, and CO_2 reacts with these groups to form carbamino compounds, or carbamates, in all vertebrates.

$$\text{protein—}NH_3^+ + CO_2 \rightleftharpoons \text{protein}\begin{matrix} \diagup COO^- \\ \diagdown H \end{matrix} + 2H^+ \qquad (2)$$

The extent of carbamate formation, however, like the hemoglobin oxygen-carrying capacity, varies in different vertebrates.

There are large differences in blood oxygen capacity, hemoglobin affinity, the magnitude of the Bohr shift, and erythrocytic volume in vertebrates, but it is difficult to correlate these differences simply with the mode of breathing. The erythrocytes of mammals are enucleated, whereas other vertebrates have nucleated cells. There is, however, a report of nonnucleated erythrocytes in fish (Hansen and Wingstrand 1960). Animals that respire in both water and air tend to have the largest erythrocytes, yet there is a considerable range of erythrocytic volumes in fish (Gulliver 1875). The question of why amphibians, like *Amphiuma,* should package their hemoglobin in such large cells is interesting but unanswered.

Relatively little is known of the details of erythrocytic metabolism except for a few mammalian species and the amphibian *Amphiuma.* What is clear is that the substrate varies as does the metabolic pathway found within the red blood cell. Trout erythrocytes, for instance, have a low glucose permeability (Bolis and Luly 1972) similar to that of *Arapaima,* whereas lungfish erythrocytes are permeable to glucose (Kim and Isaacks 1978). Organic phosphates are in high concentration in erythrocytes; these compounds bind to hemoglobin and reduce hemoglobin–oxygen affinity. The type of organic phosphate varies, being 2,3-diphosphoglycerate in most mammals and the armored catfish (Isaacks et al. 1978). Guanosine triphosphate (GTP) and adenosine triphosphate (ATP) are more typical of fish. Inositol pentaphosphate, previously reported in birds and turtles, has been found in erythrocytes of the air-breathing fish *Arapaima,* and uridine mononucleatides and inositol diphosphate as well as ATP and GTP have been reported for lungfish erythrocytes (Bartlett 1978a,b,c). The reason for these differences in organic phosphates found in erythrocytes is not clear, nor does there appear to be any clear separation between water- and air-breathing vertebrates. It must be emphasized, however, that erythrocyte metabolism is only described for cells from a few vertebrate species.

There is a great deal of variability in the properties of hemoglobin in fish and other vertebrates (see Riggs 1979 for references). It has been concluded that air-breathing fish tend to have a larger blood oxygen capacity, a lower hemoglobin–oxygen affinity, and a larger Bohr shift than water-breathing fish (Johansen et al. 1978). There is, however, considerable variability with regard to these factors in both water- and air-breathing fish; and if large numbers of fish are compared, these generalizations are not tenable (Powers et al. 1979). If, however, closely related species of air-breathing and water-breathing fish are compared (e.g., *Hoplerythrinus* with *Hoplias*), then the conclusions of Johansen et al. (1978) hold true. The air-breathing species do have a lower hemoglobin–oxygen affinity, a higher blood oxygen capacity, and a larger Bohr shift. There is also a tendency for organic phosphate levels in the erythrocytes to be higher in air-breathing fish. These differences probably reflect the increased oxygen availability to air-breathing forms.

A decrease in pH not only causes a decrease in oxygen affinity (Bohr shift) but also a decrease in blood oxygen capacity (Root shift) in most fish. Generally the magnitudes of the Bohr and Root shift are similar and appear to be due to similar processes at the molecular level. Those air-breathing fish with a large Bohr shift also have a large Root shift (Powers et al. 1979). Carter and Beadle (1931) hypothesized that, because of their elevated CO_2 levels, air-breathing fish would have a reduced Root shift. They supposed that if there was a marked Root shift, these high CO_2 levels would cause a fall in blood pH and hence a marked reduction in blood oxygen content. Blood pH, however, is similar in air- and water-breathing vertebrates independent of the blood CO_2 level (see Figure 2.16).

Carbon dioxide affects hemoglobin–oxygen affinity not only by its action on pH but also by reacting with NH_2 terminals on hemoglobin to form carbamates. Little is known about carbamate formation in fish hemoglobin, but there appears to be less carbamate formation than in mammals. Fish hemoglobin α-chains probably have N-acetylated amino termini, and so there is no carbamate formation on the α-chain (Riggs 1979). Carbon dioxide does have an effect, however, on hemoglobin–oxygen affinity that is independent of pH; this effect is probably due to carbamate formation on the β-chain (Farmer 1979). Organic phosphates compete with CO_2 for the amino groups on the β-chain, and in the presence of such compounds carbamate formation may be very small, especially in vivo where blood CO_2 levels in fish are only of the order of a few millimeters of mercury. Thus carbamate formation may be negligible in fish, first because of the restricted number of available amino termini, and second because of competitive binding by organic phosphates at the low CO_2 levels found in vivo in fish. There appears to be little difference in carbamate formation between water- and air-breathing fish (Farmer 1979) as judged by the effect of CO_2 on hemoglobin–oxygen affinity at constant pH. Carbon dioxide levels tend to be higher in air-breathing forms, but then so are the levels of organic phosphates. Air-breathing fish retain an essentially aquatic pattern of CO_2 excretion (see next section); so it is perhaps not surprising that they retain CO_2–hemoglobin binding properties similar to those found in water-breathing animals. Carbamates are estimated to contribute about 10–12% of the CO_2 exchange in mammals (Rossi-Bernardi and Roughton 1970), but this high value may be typical only of vertebrates like reptiles, birds, and mammals.

Carbon dioxide excretion and hydrogen ion regulation

It is through bicarbonate and carbamate formation that changes in CO_2 result in marked changes in $[H^+]$ in body fluids. The changes in $[H^+]$ have a marked effect on the dissociation and configuration of proteins, and these changes are usually detrimental to the organism. Hence, in the face of a continuous and variable production, CO_2 excretion must be regulated to prevent oscillations in body $[H^+]$ and the subsequent disruption of metabolism. We emphasize that, because molecular CO_2 is not very toxic, CO_2 excretion is regulated to control body pH and not the CO_2 content of the body. Similarly, the body CO_2 content can also be varied to prevent pH changes due to other agents like lactic acid. A reduction in molecular CO_2 levels will tend to raise pH, whereas CO_2 retention or the excretion of bicarbonate will lower pH.

Carbon dioxide and bicarbonate ions are small molecules and diffuse rapidly in aqueous solutions. The movement of CO_2 is probably facilitated within cells owing to the codiffusion of bicarbonate. This is of importance, however, only in the presence of carbonic anhydrase and proteins. The proteins are necessary to increase the movement of protons, which diffuse in protein solutions at rates some 1000 times greater than in water (Gros and Moll 1974), and carbonic anhydrase catalyzes the rapid interconversion of CO_2 bicarbonate pools. The importance of the codiffusion of bicarbonate in facilitating CO_2 movement increases with elevated pH and decreasing CO_2 levels.

Carbon dioxide moves easily through cell membranes which are relatively impermeable to charged particles such as HCO_3^- or H^+. There is a rapid equilibration of CO_2, with essentially no limitation on the movement of molecular CO_2 across cell membranes. In general, however, the movements of HCO_3^- and H^+ are restricted and regulated across cell membranes, bicarbonate being exchanged for chloride, and sodium for hydrogen (Figure 2.5). These mechanisms, located in the plasma membrane, are utilized to regulate intracellular pH and, coupled with the high concentration of buffers and the regulation of cation levels via plasma membrane transport, reduce the oscillations in intracellular $[H^+]$ in the face of varying rates of CO_2 and acid production.

At body pH, most of the CO_2 is stored as bicarbonate but crosses membranes as molecular CO_2. The uncatalyzed interconversion of HCO_3^- and CO_2 is very slow (Figure 2.6), especially at low temperatures. At 37°C the half-time for the uncatalyzed reaction is probably about 5–9 sec in mammalian plasma (Forster and Crandall 1975), whereas at 10°C the half-time, by extrapolation, is probably about a minute. The enzyme carbonic anhydrase catalyzes these reactions and is present in most cells, but is absent in extracellular spaces. Thus, within cells interconversion of CO_2 and bicarbonate is rapid, whereas in extracellular fluids uncatalyzed interconversion is a slow process. A sudden drop in extracellular P_{CO_2} will cause a rapid movement of CO_2 out of the cell, and the catalyzed dehydration of intracellular bicarbonate will occur much more rapidly than extracellular bicarbonate dehydration. Thus, under these conditions, extracellular bicarbonate will change much more slowly than intracellular bicarbonate. The mammalian red blood cell (and probably bird and reptile red blood cells) is very permeable to bicarbonate, such that when P_{CO_2} of the blood is reduced

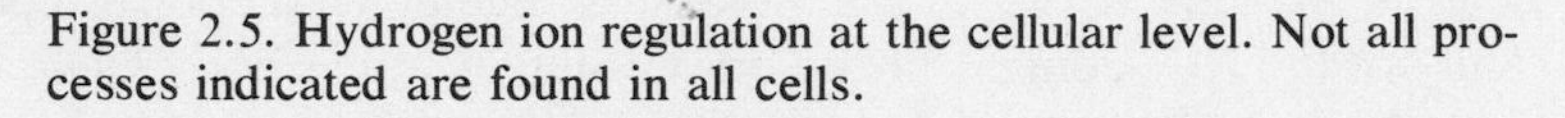
Figure 2.5. Hydrogen ion regulation at the cellular level. Not all processes indicated are found in all cells.

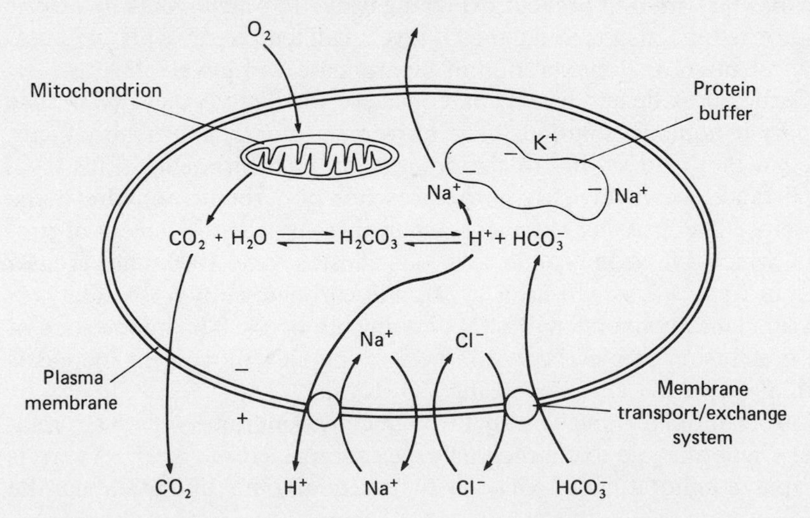

and erythrocytic bicarbonate falls, plasma bicarbonate moves into the erythrocyte and is converted to CO_2 at the catalyzed rate. Mammal, bird, and reptile blood plasma, therefore, shows rapid oscillations in bicarbonate levels closely correlated to changes in P_{CO_2}. There are probably small time lags (milliseconds) between the oscillations in P_{CO_2} and intracellular and extracellular bicarbonate, related to the movement of bicarbonate into the erythrocyte (Hill et al. 1977). If extracellular bicarbonate could not enter the red blood cell and form CO_2 at the catalyzed rate, then the intracellular and extracellular oscillations in bicarbonate could be several seconds out of phase.

Mammals excrete most of the CO_2 as molecular CO_2 into the lung gas, although there is also a low rate of bicarbonate excretion via the kidney. Most of the CO_2 is transported as bicarbonate in plasma. At the lungs bicarbonate enters the erythrocytes in exchange for chloride and is rapidly dehydrated to CO_2, which diffuses into the alveolar gas (Figure 2.7). The lung epithelium of the rabbit is also capable of dehydrating plasma bicarbonate (Effros et al. 1978); however, the relative importance of this pathway for CO_2 excretion in mammals has not been ascertained.

The pattern of CO_2 excretion across the gills of fish is somewhat different from that in mammals. It appears that many, if not all, aquatic animals regulate bicarbonate rather than molecular CO_2 levels in the face of a pH shift (Cameron 1978a). High CO_2 levels do cause changes in gill ventilation in fish, but the increase in water flow has little effect on either blood P_{CO_2} or pH (Janssen and Randall 1975). The trout

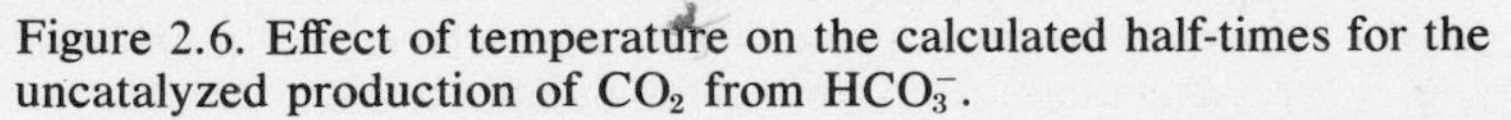

Figure 2.6. Effect of temperature on the calculated half-times for the uncatalyzed production of CO_2 from HCO_3^-.

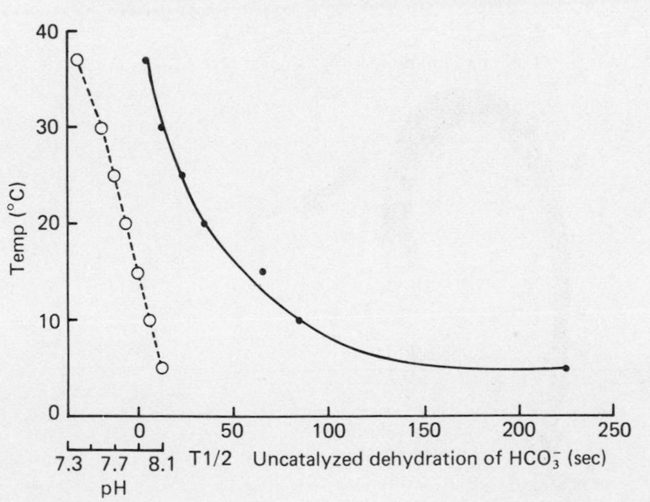

raises plasma bicarbonate to adjust pH during hypercapnia, and it has been shown that the dogfish is able to take up bicarbonate directly from the water passing over the gills (Randall et al. 1976). This indicates that, unlike the situation in mammals, manipulation of molecular CO_2 via changes in ventilation is not utilized to regulate pH.

It is not clear what role ventilation plays in determining the rate of CO_2 excretion in fish. Certainly a sharp reduction in ventilation such as occurs in fish during hyperoxia (Dejours 1973; Randall and Jones 1973) causes a marked rise in blood CO_2 levels. Also an increase in ventilation such as occurs during hypoxia can result in a slight rise in blood pH, presumably as a result of a fall in CO_2 levels in the blood (see Figure 5.1). Thus changes in ventilation affect blood CO_2 levels, but it appears that fish do not regulate ventilation in order to adjust P_{aCO_2} and therefore pH. These animals live in an environment containing relatively little oxygen, and ventilation is adjusted to maintain oxygen delivery to the gills (Randall and Cameron 1973).

What then is the pattern of CO_2 excretion in fish, and how is it regulated? Most CO_2 is excreted across the gills with only a small amount being lost via the kidney, even in freshwater fish with high urine flow rates. The gill epithelium contains high levels of carbonic anhydrase, and plasma bicarbonate enters the gill epithelium, along with H^+, and is dehydrated to CO_2, which diffuses into the water (Haswell et al. 1980). It seems likely that the outer membrane of the gill epithelium is relatively impermeable to both HCO_3^- and H^+ and, in

Figure 2.7. Excretion of plasma bicarbonate in mammals, which occurs via the erythrocyte (RBC) where bicarbonate dehydration is catalyzed by the enzyme carbonic anhydrase.

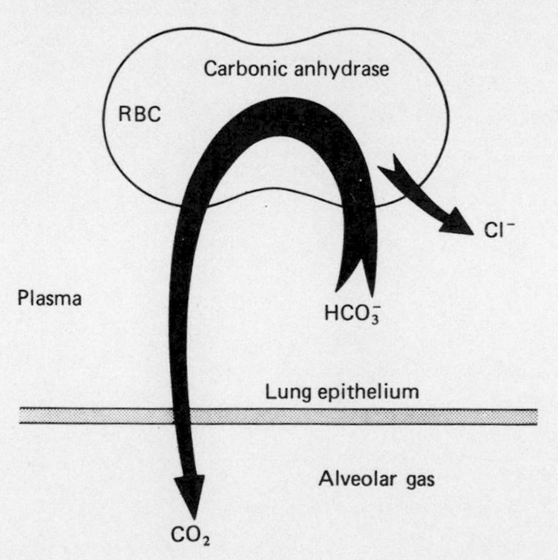

freshwater fish at least, there are HCO_3^-/Cl^- and Na^+/H^+ exchange mechanisms (Maetz and Garcia-Romeu 1964) on the surface (Figure 2.8). Cation and anion exchange probably constitute only a minor portion of total CO_2 excretion (Cameron 1976). It is also possible that some H^+ is exchanged for Na^+ and bicarbonate for chloride on the blood side of the gill epithelium.

In seawater fish the anion and cation exchange mechanisms on the external border of the gill epithelium are probably reversed, the bicarbonate and hydrogen ions taken up from seawater forming CO_2 that diffuses back into seawater leaving behind a water molecule for each bicarbonate ion taken up (Figure 2.9). This process will tend to reduce water loss by hydrating the gill epithelium.

The evidence for bicarbonate excretion via the gill epithelium is based on experiments carried out on rainbow trout, and is as follows: (1) Ninety percent of the blood total CO_2 is present in plasma, and 95% of the plasma CO_2 pool is as bicarbonate. Thus plasma bicarbonate is the major form of CO_2 in blood, and there is a marked reduction in plasma bicarbonate between blood entering and leaving the gills. (2) Blood transit time through the lamellae is about 2 sec, too rapid for the uncatalyzed dehydration reaction. (3) Inhibition of carbonic anhydrase with acetazolamide (Diamox) reduces CO_2 excretion, indicating that the catalyzed reaction is required to excrete bicarbonate. (4) The gill epithelium contains high levels of carbonic anhydrase, and the isolated, perfused, erythrocyte-free gill preparation will excrete HCO_3^- from a perfusate even though transit time is of the order of only a second. The CO_2 excretion is blocked by carbonic anhydrase inhibition, indicating that the bicarbonate is converted to CO_2 and then excreted (Haswell and Randall 1978).

The erythrocytes contain high levels of carbonic anhydrase and the question can be asked, why does plasma HCO_3^- enter the gill epithelium

Figure 2.8. Pattern of CO_2, hydrogen ion, and other ion movement through the teleost gill epithelium in freshwater.

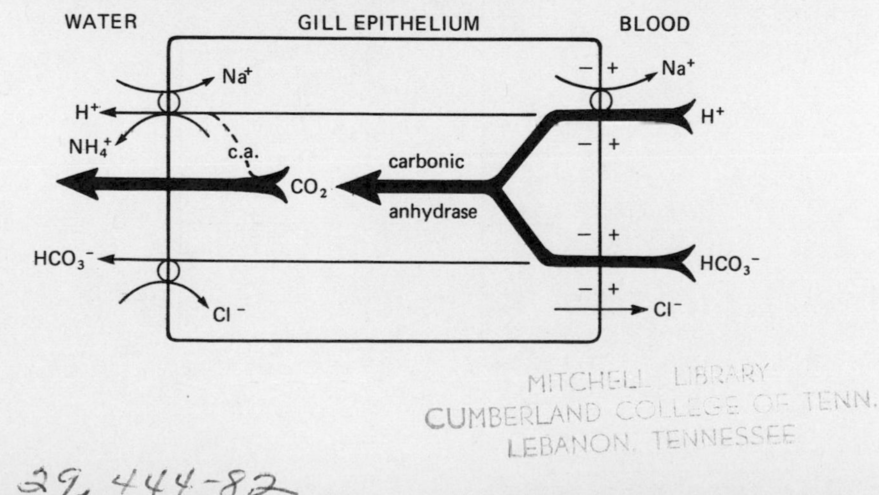

in preference to the erythrocyte? The answer to this question is that (1) the enzyme substrate affinity of gill epithelial carbonic anhydrase is higher than that found in the erythrocytes (Haswell 1978), and (2) Haswell and Randall (1976) showed that intact erythrocytes in plasma did not catalyze plasma bicarbonate in vitro, whereas intact erythrocytes in Cortland saline did catalyze plasma bicarbonate dehydration. Hemolyzed cells in plasma, however, catalyzed the formation of CO_2. It was concluded that something in the plasma reduced the permeability of erythrocytes to plasma bicarbonate. The method used, however, has some technical limitations and the red blood cells may be bicarbonate permeable because teleost erythrocytes are chloride-permeable (Haswell et al. 1978) and show a chloride shift when exposed to elevated CO_2 levels (Cameron 1978b). These observations are difficult to reconcile at this time especially in the face of the observation that dogfish erythrocytes are permeable to bicarbonate ions (Obaid et al. 1980).

The erythrocytes do contain high levels of carbonic anhydrase, and increases in P_{CO_2} levels result in the rapid formation of bicarbonate within the erythrocytes. Fish erythrocytes swell rapidly under these conditions owing to water entry, presumably as a result of the rapid increase in intracellular bicarbonate levels. Within the tissue nearly all

Figure 2.9. The pattern of CO_2 and ion movement across the gill epithelium in seawater. Note that the uptake of bicarbonate and hydrogen ions results in the uptake of water into the gill epithelium following the formation of CO_2.

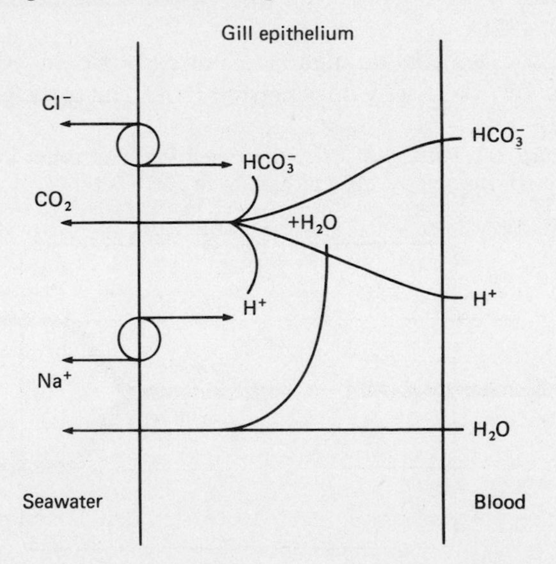

the bicarbonate formed as a result of CO_2 entry from metabolizing tissues will be situated in the red blood cell (RBC). If the bicarbonate is retained in the erythrocytes, then CO_2 can be unloaded at the gills by catalysis of RBC bicarbonate pools. Circulation time in fish is long compared with mammals, being of the order of a minute, and blood is probably resident in veins for the order of 40–50 sec, enough time for the uncatalyzed reaction in plasma to proceed toward equilibrium. Blood leaving the tissues will have elevated CO_2 and erythrocytic bicarbonate levels. The CO_2–bicarbonate system in the plasma may not be in equilibrium; CO_2 levels will be high. In the veins the uncatalyzed reaction will occur, raising plasma bicarbonate levels. The blood within the veins is in a closed system; so a rise in plasma bicarbonate can only occur at the expense of plasma CO_2 levels. This will lead to a reduction in erythrocytic CO_2 and therefore HCO_3^- levels. Thus in the veins erythrocytic HCO_3^- will be reduced and plasma HCO_3^- increased, associated with a gradual fall in plasma pH but a rise in erythrocyte pH. The latter will result in the binding of O_2 by hemoglobin and a lowering of blood P_{O_2}. Such a change was observed in blood leaving the eel swim bladder rete (Berg and Steen 1968), and a slow "Bohr on" shift was observed in eel erythrocytes placed in an alkaline environment (Forster and Steen 1969). Both of these observations are consistent with reduced erythrocytic bicarbonate permeability in fish.

Thus, CO_2 entering the blood from tissues rapidly forms bicarbonate within the erythrocytes. In the plasma, CO_2 hydration proceeds at the uncatalyzed rate, and in the veins CO_2 levels fall as bicarbonate levels in the plasma rise. Erythrocytic bicarbonate is dehydrated, and CO_2 moves from the RBC into the plasma, the rate being determined by the uncatalyzed hydration reaction within the plasma (Figure 2.10). At the gills, CO_2 diffuses from the blood, and erythrocytic bicarbonate is rapidly dehydrated; however, plasma bicarbonate enters the epithelium (not the RBC) and is either dehydrated to CO_2 or exchanged for chloride at the water interface.

The gill is a hydrogen ion–excreting tissue, with the regulation of plasma acid–base status intimately linked to cationic and anionic exchanges occurring across the gill (De Renzis and Maetz 1973; Haswell et al. 1980). Proton pumping and its relationship to CO_2 excretion and salt movements in the trout are summarized in Figure 2.8. By controlling epithelial cell pH, the movements of plasma bicarbonate and hence plasma pH can be modulated in fish. Epithelial cell pH is modulated principally by cationic, and to a lesser extent anionic, exchanges occurring across the apical membrane (water–gill interface) of the gill (Haswell et al., 1980). An increased rate of Na^+/H^+ exchange would result in an elevation of cytoplasmic pH. This increase in cytoplasmic pH will shift Equation (1) to the right, resulting in a build-up of bicarbonate within the cell. The rise in cytoplasmic HCO_3^- will elevate plasma bicarbonate levels by reducing the gradient for bicarbonate entry at the blood–gill membrane. In a similar fashion a reduction of sodium uptake rates would result in a fall in plasma bicarbonate levels. Proton pumping across the fish gill is remarkably similar to that in the amphibian and turtle urinary bladder, a tissue also capable of hydrogen-ion pumping. In both tissues proton pumping is coupled to CO_2 and salt movements; but in the gill hydrogen ions are "dumped" into the external environment, whereas in the bladder protons are secreted into an internal storage and/or excretion site, the urine contained within the bladder.

The pattern of CO_2 excretion is, therefore, different in water-breathing fish and air-breathing mammals. The pattern of CO_2 excretion in fish enables them to regulate plasma HCO_3^- and therefore plasma pH (Randall and Cameron 1973). It seems probable that blood

Figure 2.10. The pattern of CO_2 movement between tissues and water via the blood in teleost fish. The $CO_2 \rightleftharpoons HCO_3^-$ reaction is catalyzed by carbonic anhydrase within the erythrocyte and gill epithelium. The reaction in the plasma occurs at the slow uncatalyzed rate.

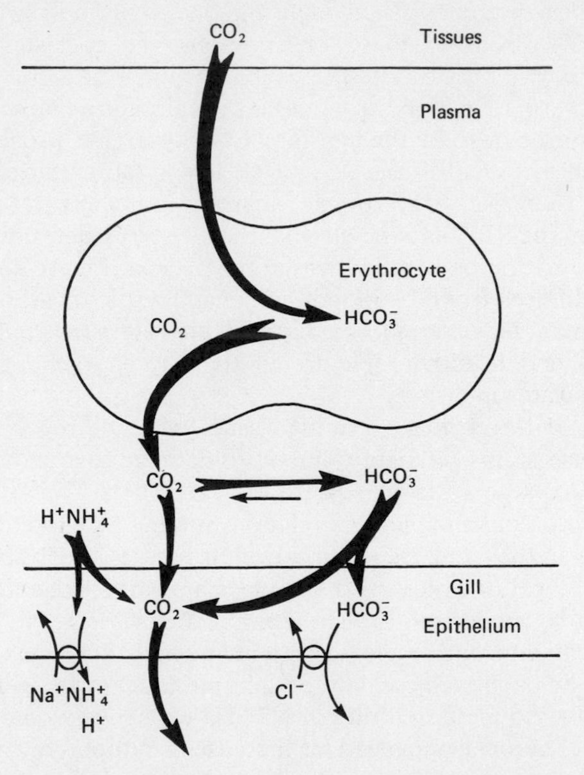

pH is controlled by regulating the entry of plasma HCO_3^- into the gill epithelium (Haswell et al. 1980).

Carbonic anhydrase is present in the erythrocytes of fish to generate protons for a rapid Bohr and Root shift driving O_2 from hemoglobin. The Bohr and Root shifts are reversed in the veins, lowering venous P_{O_2} and thereby augmenting O_2 uptake at the gills. Formation of carbamino compounds in the erythrocytes will reduce the oscillations in bicarbonate and, therefore, the capacity to regulate CO_2 excretion via the gill epithelium. Thus carbamino formation is minimal and may even be selected against in fish and some amphibians. The pattern of CO_2 excretion in fish is designed not to excrete CO_2 per se but to couple CO_2 excretion to ion and H^+ transport and therefore regulate body $[H^+]$.

Air breathing in fish

Some fish are facultative air breathers during particular stages of their life cycle, or during certain seasons, or under a given set of environmental conditions, but are obligatory air breathers under other conditions. Adult *Arapaima* cannot survive if denied access to air even in well-aerated water (Stevens and Holeton 1978). The gar *Lepisosteus* is a facultative air breather at low temperatures but becomes an obligate air breather when O_2 uptake increases at high temperatures (Rahn et al. 1971). *Piabucina* obtains 10% of O_2 from air if water P_{O_2} is high, but 70% of oxygen uptake from air when water P_{O_2} is 35 mmHg (Graham et al. 1977). Young forms of *Anabas, Clarius, Heteropneustes,* and lungfish rely less on aerial breathing than the adult forms. Thus the extent of air breathing varies both within and between fish species (Singh and Hughes 1971; Johansen et al. 1970; Rahn et al. 1971). Bimodally breathing fish are obligatory air-breathing forms only for oxygen uptake and must ventilate their gills for CO_2 excretion and $[H^+]$ regulation; that is, they are obligatory water-breathing forms for acid–base regulation and obligatory air breathers for oxygen uptake.

Gills are, in general, not suitable organs for gas exchange in air because the lamellae on them collapse and stick together, and water is retained in the small spaces between lamellae, these events producing a tremendous reduction in functional respiratory surface area available for gas exchange. A few species have increased structural support of their gills and have widely spaced lamellae, such that their gills can be used for gas exchange in air, as, for example, *Mnierpes* (Graham 1973). These fish are the exception, however, and most air-breathing vertebrates utilize organs other than gills for air breathing. These organs include the fish gas bladder, a dorsal pouchlike outgrowth of the anterior region of the gut, and the lung, which is similar to the gas bladder of fish in general form, but which opens ventrally from the pharynx, rather than having a dorsal location (Figure 2.11). Many fish use elaborations of membranes in the buccal, opercular, or pharyngeal cavity, the gut, or the gas bladder for air breathing (Figure 2.12). Some fish, many amphibians, and some snakes and other reptiles use the skin surface for gas exchange. For example, *Mnierpes* (Graham 1973) and the eel *Anguilla* (Berg and Steen 1966), use their skin for gas exchange in air. The extent of development of the accessory breathing organ is correlated with the extent to which the fish utilizes air breathing for an oxygen supply. It would, of course, be surprising if the system were other than conservative, and the breathing organ had a larger gas transport capacity than required.

Those animals that use the same structure, namely the gills, to obtain oxygen from both water and air cannot do so simultaneously. Many of these animals, like the mud skipper *Periopthalmus* and *Mnierpes,* are intertidal animals that spend short periods of time out of water, and are exceptional in that they are often leaving a relatively well-oxygenated

Figure 2.11. Diagrammatic cross sections and longitudinal sections of the air bladders of various fish and tetrapods, showing variations in the position of the opening, the number of lobes, and the extent of internal development (from Romer 1970).

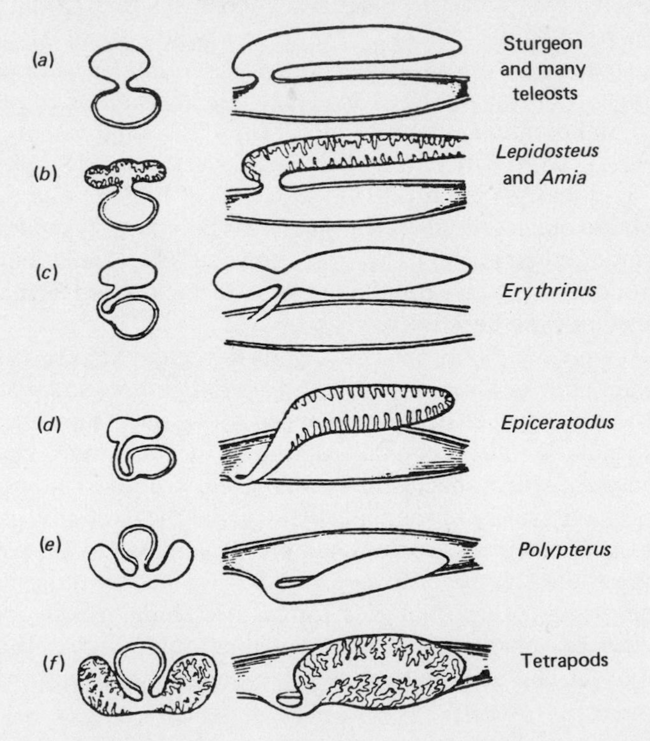

aquatic environment. This group of air-breathing fish is different from other species in that if the gills or structures modified from gills are used for air breathing, normal rates of CO_2 excretion are maintained during air exposure (Graham 1973). In those air-breathing fish with air-breathing organs not derived from gills, very little CO_2 is released into air, and most is excreted across the gills, which are usually bathed with water.

Gills, if used for gas transfer in air, can only be used for relatively short periods (Graham 1973), and more permanent terrestrial vertebrates that use only gills in air breathing have not evolved. Gills are also used for ion exchange and acid–base regulation, as well as ammonia excretion and gas transfer. Although normal rates of O_2 and CO_2 transfer can be maintained across the gills of some fish in air (Figure 2.13), it is clear that the gills cannot function normally in NH_4^+ excretion unless the animal is in water. Most fish have the capacity to produce some urea, which is less toxic than ammonia. It has been shown that *Periopthalmus* has an enhanced capacity to produce urea compared with totally aquatic fish (Gregory 1977), presumably maintaining low levels of ammonia in the body during air exposure.

A more common occurrence is for fish to obtain oxygen from air via an accessory breathing organ while continuing to ventilate the gills with water. This enables the fish to maintain ion exchange and ammonia excretion while obtaining oxygen from air. This can create problems, however, because the accessory breathing organ has a blood supply in parallel with the systemic circuit, and oxygenated blood from the air-breathing organ is pumped through the gills before reaching the body tissues. Often air breathing is in response to aquatic hypoxia, and oxygenated blood from the air-breathing organ could lose oxygen to the water across the gills. Clearly fish do not want to expend energy in air breathing in order to oxygenate the water. Thus in these fish there are strong selective pressures favoring a reduction of the gills to prevent oxygen loss to the water. A definite trend of reducing gill surface area and/or increasing lamellar diffusion distance is evident in air-breathing teleosts. Figure 2.14 illustrates secondary lamellae from the trout (completely aquatic teleost), the jeju (facultative air breather), and the pirarucu (obligate air breather), and shows the marked changes associated with air breathing. The overall effect is a marked reduction in the gill diffusing capacity with the trend toward air breathing. In some instances there are wide basal channels connecting afferent and efferent filament vessels (Figure 2.15b), which run deep in the body of the filament and essentially form a lamellar bypass channel because of the large diffusion barrier between blood and water. In the lungfish, *Lepidosiren* and *Protopterus,* there is a total loss of lamellae on certain gill arches, and these vessels form direct shunt connections between

Figure 2.12. The gas bladder of teleost fish: (*a–d*) from the obligate air-breathing fish, the pirarucu, *Arapaima gigas*; (*e–h*) from the facultative air-breathing fish, the jeju, *Hoplerythrinus unitaeniatus*. (*a*) The pirarucu. (*b*) Ventral incision showing the extensive respiratory surface.

(*a*)

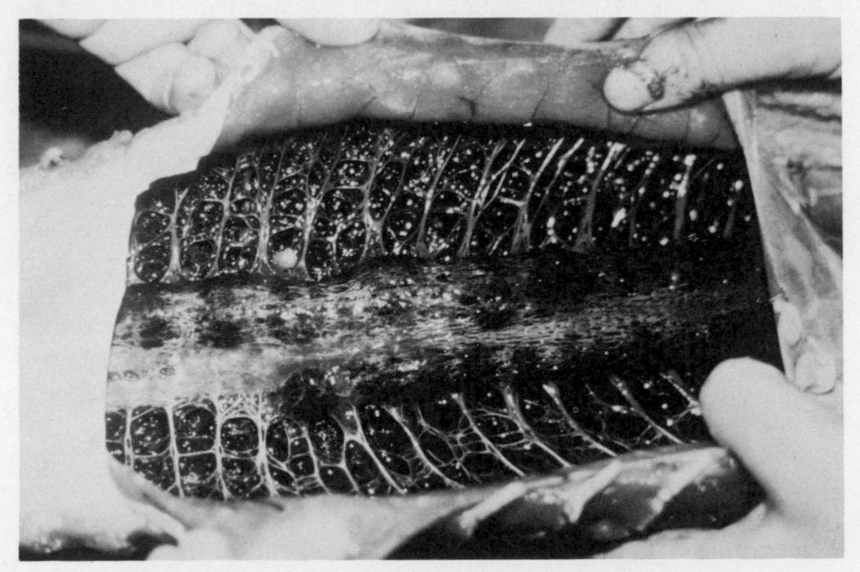

(*b*)

ventral and dorsal aortae (Figure 3.10). All these modifications reduce oxygen loss across the gills.

Oxygen and CO_2 diffuse at similar rates; thus a reduction in gill area will also affect CO_2 excretion. Larger CO_2 gradients are required to maintain CO_2 excretion, and generally CO_2 levels are elevated in air-breathing fish, reflecting the reduced diffusing capacity of the gills. Arterial P_{CO_2} in trout is about 2 mmHg, in jeju 10–12 mmHg, in pirarucu 28 mmHg (Randall et al. 1978a,b). Carbon dioxide is not toxic

Figure 2.12. (*c*) A plastic cast of the capillary circulation to the bladder air sacs. (*d*) Section through a single small air sac. Continued.

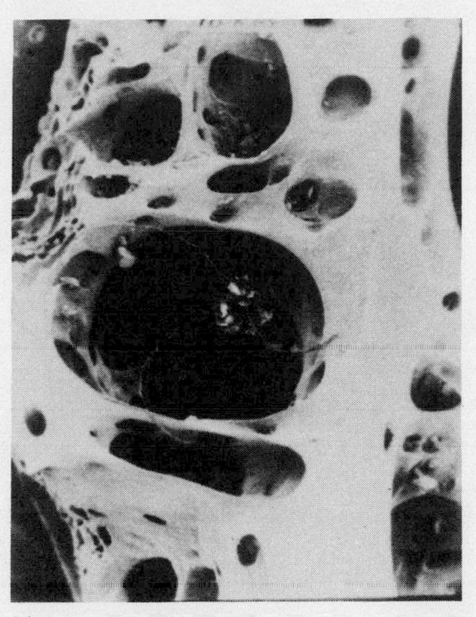

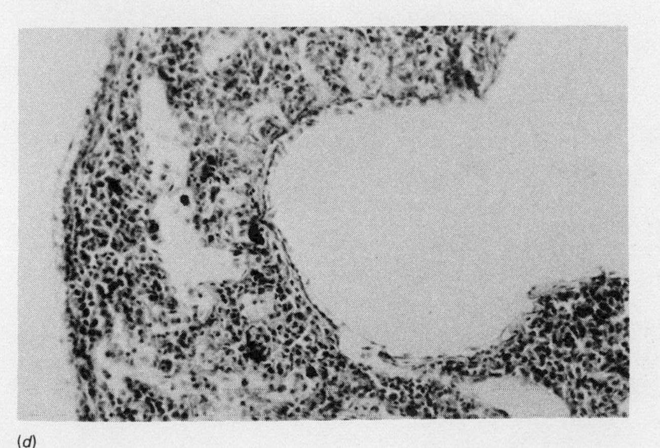

(*d*)

at these levels as long as pH is adjusted to values appropriate for that temperature (Reeves et al. 1977); thus bicarbonate levels are also elevated in air-breathing animals (Figure 2.16).

The gourami, *Trichogaster,* an anabantid, has an aerial gas exchange organ, the labyrinth, which is embryonically derived from gill tissue. These fish also retain gills for water breathing. The circulation to the water- and air-breathing organs is arranged in series (Figure 2.17). Blood returning to the heart is pumped to the body tissues via the third

Figure 2.12. (*cont.*) (*e*) The jeju. (*f*) An isolated and opened gas bladder; note only a portion of the gas bladder is vascularized.

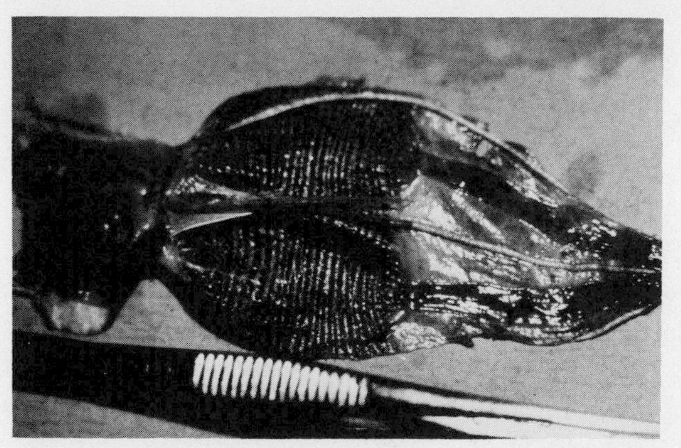

(*f*)

and fourth gill arches, which are reduced in size. Under hypoxic conditions in water (probably a common occurrence in the gourami's natural habitat), CO_2 is excreted across the gills, and O_2 obtained from air in the labyrinth. It seems probable that there is separation of oxygenated and deoxygenated blood in the heart and ventral aorta, though this has yet to be investigated. Preferential flow of oxygenated blood from the labyrinth to gill arches three and four and thence the body would prevent O_2 loss into hypoxic water across the well-developed gill arches

Figure 2.12. (*g*) Plastic cast of the microcirculation to the vascular portion of the bladder. (*h*) Transverse section through air sacs in the bladder. (*c, d, g,* and *h* supplied by Dr. B. Gannon; *e* and *f* by Dr. G. Holeton.)

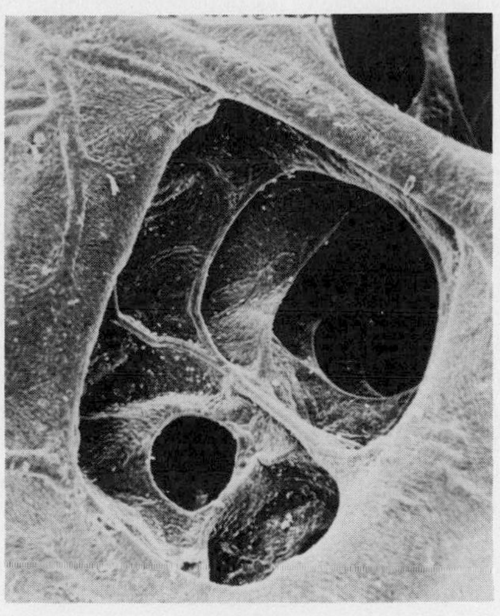

(*g*)

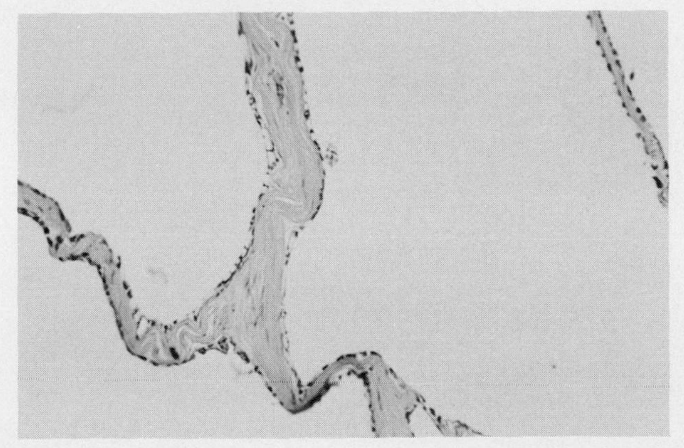

(*h*)

one and two, which would receive deoxygenated blood from the body. Perfusion of gill arches one and two is required for CO_2 excretion into water; however, this fish when exposed in air, can maintain the same rate of CO_2 excretion across the labyrinth into air because the epithelium, derived from gill tissue, contains high levels of carbonic anhydrase.

Carbon dioxide excretion and acid–base regulation in bimodal-breathing vertebrates

Although the gills are reduced in size in air-breathing fish, most of the CO_2 is still excreted across the gills rather than the air-breathing organ. Why are the gills retained for CO_2 excretion in air-breathing fish? Why not excrete CO_2 via the air-breathing organ? Carbon dioxide excretion into air would reduce the requirement for CO_2 excretion via the gills; however, we emphasize that in fish the pattern of CO_2 excretion is designed to retain control of HCO_3^- movements across the gills and, therefore, body pH. Carbon dioxide loss via the air-breathing organ would reduce the capacity of the animal to regulate pH via HCO_3^- excretion across the gills.

In general, the excretion of CO_2 via the aerial exchange organ is low,

Figure 2.13. *Mnierpes macrocephalus*. Influence of weight on rates of aerial and aquatic oxygen consumption and aerial CO_2 evolution. Solid circles, upper line: aerial oxygen consumption; open circles, dashed line: aquatic oxygen consumption; triangles, lower line: aerial CO_2 evolution (from Graham 1973).

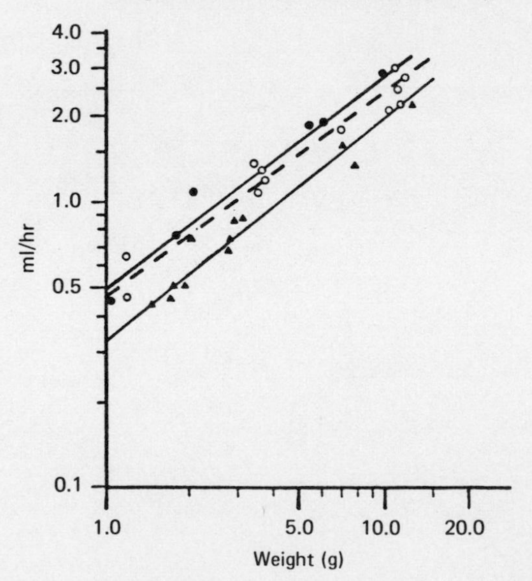

Figure 2.14. Plastic casts of a portion of the gill filament of: (*a*) rainbow trout (*Salmo gairdneri*); (*b*) jeju (*Hoplerythrinus unitaeniatus*); (*c*) pirarucu (*Arapaima gigas*). In (*a*) blood (b, black arrows) enters along the afferent filament artery (afa), and leaves via the efferent filament artery (efa). Also visible are the central canal (cc) with its associated afferent (acv) and efferent companion vessels (ecv). Water flow (W, white arrows) is in the opposite direction to blood flow. (Casts by Dr. B. Gannon.)

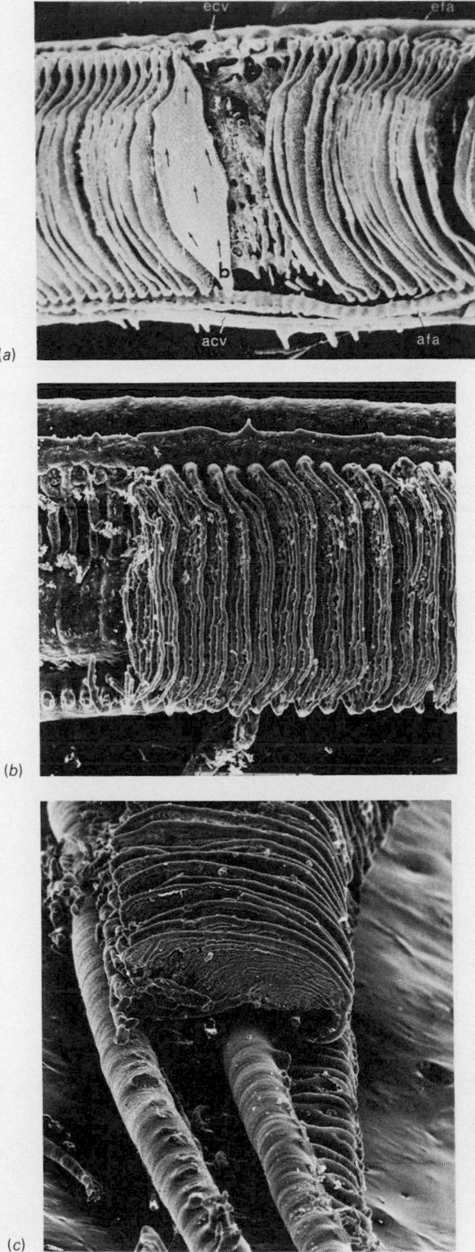

first, because the blood returning from the tissues passes first through the gills where most of the CO_2 is removed. In addition, the epithelium of the air-breathing organ does not contain carbonic anhydrase except in those organs derived from gill tissue (Burggren and Haswell 1979).

Figure 2.15. Plastic casts of the secondary lamellae of: (*a*) rainbow trout; (*b*) pirarucu. (Casts supplied by Dr. B. Gannon.)

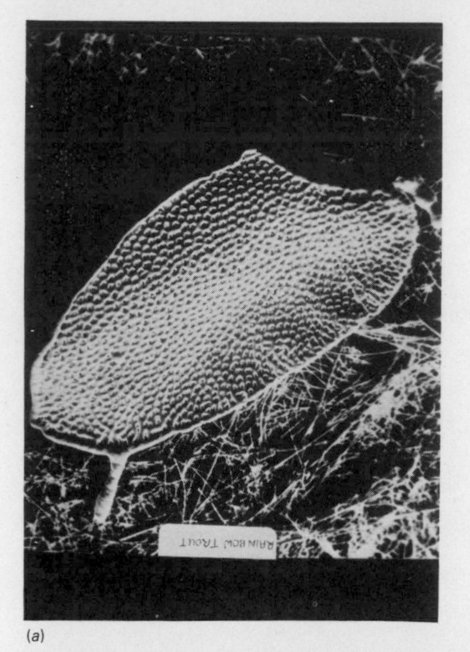

(*a*)

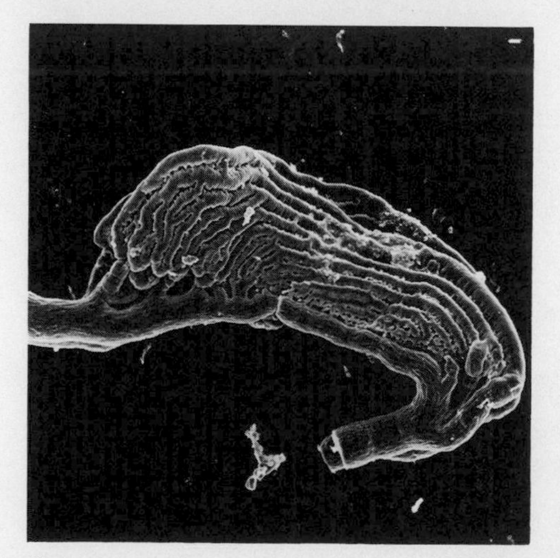

(*b*)

Thus, because fish erythrocytes have limited ability to catalyze the dehydration of plasma bicarbonate, CO_2 formed from plasma bicarbonate will be excreted into the lung at a slow rate determined in part by the uncatalyzed dehydration reaction.

If, in air-breathing fish, the gills become nonfunctional, as can occur during air exposure, CO_2 levels in the body will rise. Hence CO_2 excretion into air increases, but not sufficiently to bring the animal into a

Figure 2.16. Blood buffer lines plotted on Davenport diagram for six species of ectotherms living in Amazon basin. Normal arterial values of unanesthetized animals are shown by black dots. P_{CO_2} isopleths are calculated for a temperature of 28°C (from Rahn and Garey 1973).

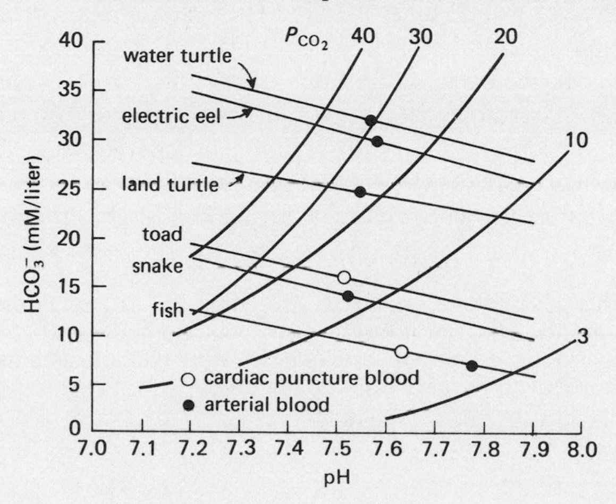

Figure 2.17. The circulation of the gourami, *Trichogaster*.

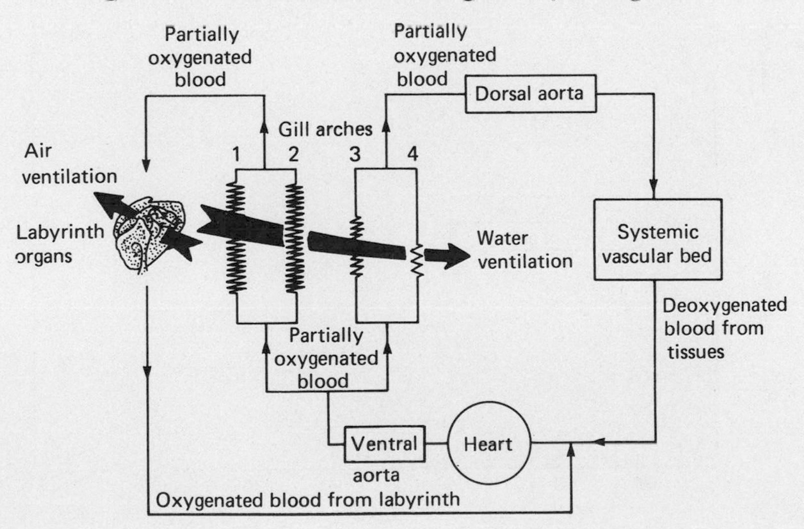

steady state (Randall et al. 1978a). Carbon dioxide excretion is still limited by the rate of bicarbonate dehydration, indicated by experiments in which the infusion of carbonic anhydrase enhanced CO_2 excretion and lowered body CO_2 levels in air-exposed fish (Figure 2.18). Thus these fish, while retaining control of CO_2 excretion (and therefore acid–base regulation) in water, suffer hypercapnia and a subsequent acidosis as well as dehydration and ion imbalance during air exposure. The pattern of CO_2 excretion found in entirely aquatic fish like trout appears to be retained in air-breathing fish in order to regulate body pH when the animal is in water. The gills are retained for ion and pH regulation, and the fish must remain in water for this process to continue. In the absence of water, hydrogen ion excretion is abolished, just as occurs when sodium uptake, which is coupled to proton excretion, is inhibited in those fish ventilating their gills with water (Kirschner et al. 1973). Therefore the transition from water to air imposes two distinct limitations on CO_2 and acid–base homeostasis. (1) Because of the change in density of the ventilatory medium (air vs. water) most gills collapse under their own weight, resulting in a large diffusion dead space. Consequently, a sufficient surface area for the loss of CO_2

Figure 2.18. Changes in dorsal aortic pH (pH_a) and P_{CO_2} (P_{aCO_2}) during air exposure, and the effect of infusion of carbonic anhydrase (C.A.) into the dorsal aorta of an individual *Hoplerythrinus* (from Randall et al. 1978a). (Reproduced by permission of the National Research Council of Canada from the *Canadian Journal of Zoology* 56: 970–3, 1978.)

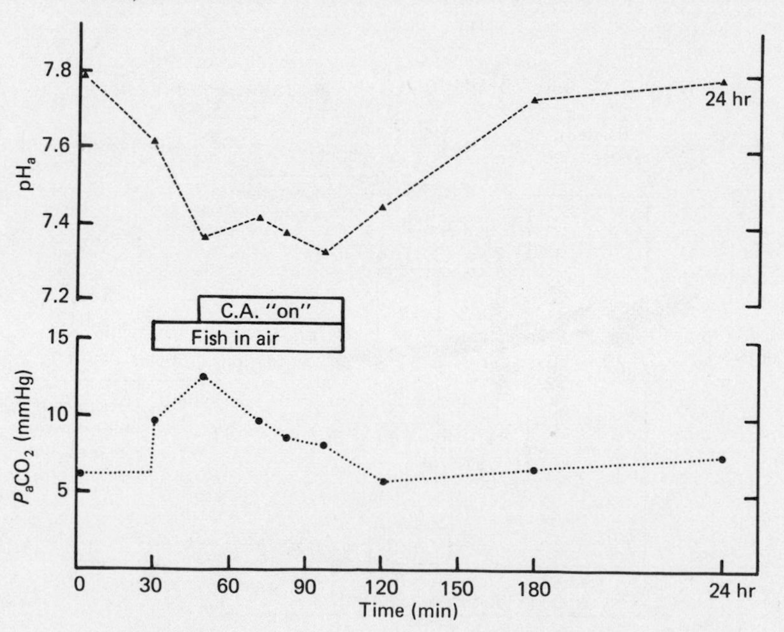

(or uptake of oxygen) is no longer available. The loss of molecular CO_2 is impeded, and a build-up of CO_2 develops. (2) In addition to this retention of CO_2, the resultant acidemia is compounded by the inability of the gill to continue proton pumping and regulate plasma bicarbonate levels, as ionic exchange is no longer possible.

Many fish use their skin for aerial gas transfer, and although subject to uncontrolled water loss, the surface is adequate for CO_2 excretion. The lungfish, *Protopterus* and *Lepidosiren,* like other fish, excrete most of their CO_2 via gills when in water. If it is exposed to moist air, the blood P_{CO_2} of *Protopterus* actually declines because skin vasodilation occurs, and CO_2 is lost via the skin (DeLaney et al. 1974). With prolonged exposure to air the lungfish retreats into a mud burrow and surrounds itself with mucus, which dries out into a cocoon. The mouth of the lungfish is connected to the outside via a narrow breathing tube. Cocoon formation results in a marked change in the breathing pattern (Figure 2.19). The tidal volume is reduced but is offset by an increase in breathing frequency such that lung ventilation does not change. Oxygen uptake nearly halves upon air exposure and falls to 20% of the aquatic level within a month. As M_{O_2} falls, blood P_{O_2} increases. At the same time blood P_{CO_2} rises to as high as 50 mmHg and is associated with a large fall in blood pH (Figure 2.20). This rise in CO_2 occurs as O_2 uptake is reduced to a low level and is not correlated with the changes in lung ventilation. It is most probably due to the loss of the skin and gills as a

Figure 2.19. Consecutive changes in mean blood pressure, heart rate, and lung breath frequency of a 3.4-kg lungfish during 283 days of artificial aestivation in a cloth sac (from DeLaney et al. 1974).

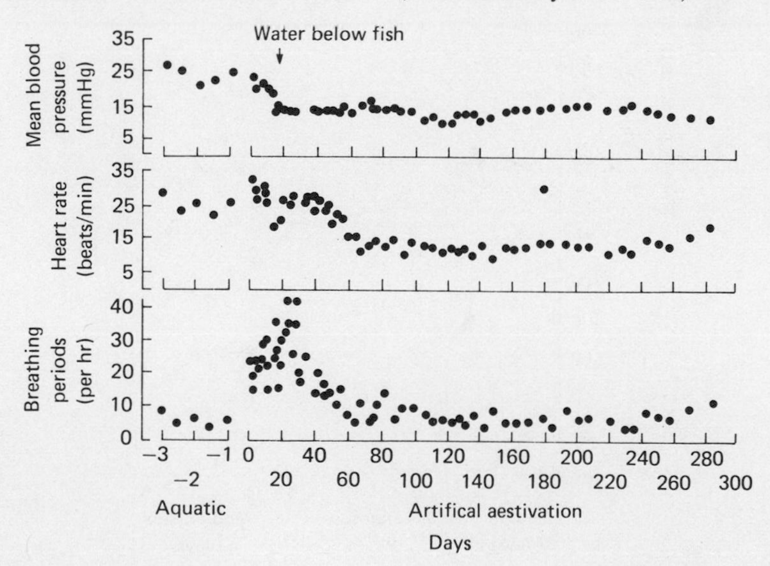

major pathway for CO_2 excretion. Thus, the lungfish resembles other fish in the qualitative nature of its response to air exposure. It has the capacity, however, to withstand this marked acidosis and accompanying dehydration for prolonged periods (Figure 2.21). During aestivation blood pH falls initially and then rises, probably in part because of Na^+/H^+ and HCO_3^-/Cl^- exchange with the tissues, HCl being dumped in the tissues. There is no urine production, and the animal functions much like a nephrectomized dog (Giebisch et al. 1955). Nothing is known of the changes in CO_2 and pH that occur when the aestivating lungfish returns to water.

What is clear is that in most bimodally breathing animals the gills and/or skin are for CO_2 excretion, and the lung or gas bladder is for oxygen uptake with normally little involvement in CO_2 excretion. This is seen in many air-breathing vertebrates, including the fish *Hopleryt-hrinus, Lepisosteus, Arapaima, Lepidosiren,* and *Protopterus* (Figure

Figure 2.20. Consecutive changes in arterial blood pH, P_{CO_2} and P_{O_2} in a nonaestivating lungfish first exposed for 20 hr to moist air and then allowed entry into mud for aestivation (from DeLaney et al. 1974).

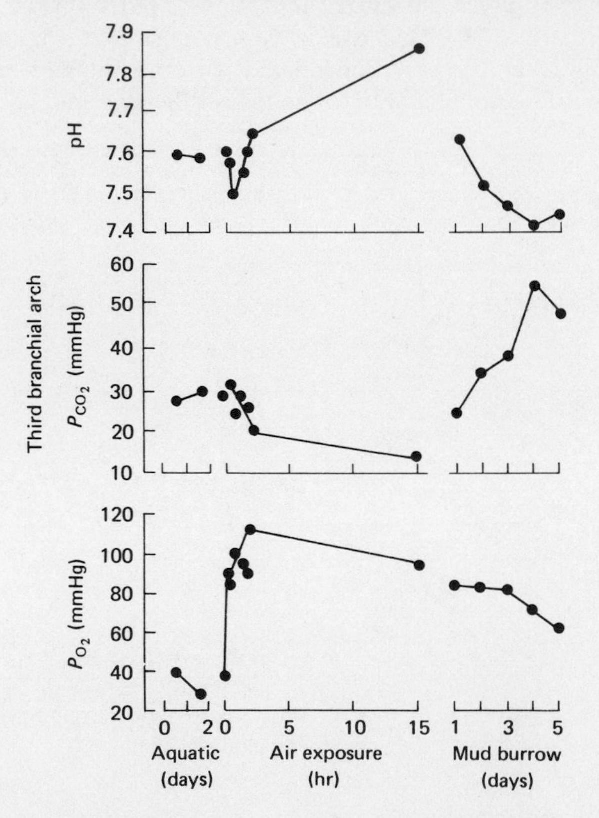

2.22) and the toad (*Bufo marinus*). Amphibia can regulate arterial P_{CO_2} in order to control blood pH (Figure 2.23). Exactly how arterial P_{CO_2} is controlled, however, is not clear. In *B. marinus* and *Xenopus laevis* in air the lung gas exchange ratio is only 0.2 (Emilio and Shelton 1974), similar to that of air-breathing fish, and most of the CO_2 is excreted via the skin. Increases in carbon dioxide cause hyperventilation in *Bufo* (Macintyre and Toews 1976), but it is not clear if blood P_{CO_2} levels are regulated by changes in lung ventilation and/or changes in skin perfusion.

In order for the total elimination of the gills to proceed in air-breathing forms, alternate sites and/or mechanisms for acid–base regulation must have evolved. In many fish (e.g., lungfish and anguilliformes) the trend towards reduction of armor in the form of scales and increased vascularization is evident, as is true for amphibians. Although this vascularization and a moist skin may provide a suitable pathway for the loss of molecular CO_2 via passive diffusion, as has been discussed, a mechanism to control hydrogen ion activity in the extracellular space is also required.

Specialized regions of skin may represent alternate sites for hydro-

Figure 2.21. Changes in acid–base status during aestivation of *P. aethiopicus*. Solid line: in vivo CO_2 titration line based on samples drawn from two lungfish living in water; dotted line: in vivo CO_2 titration line of one lungfish after 7 months of aestivation; small open circles: hyperventilation values in a single lungfish disturbed after 7 months aestivation. Breathing frequency increased from 6 to 15–18/hr in this animal (from DeLaney et al. 1977).

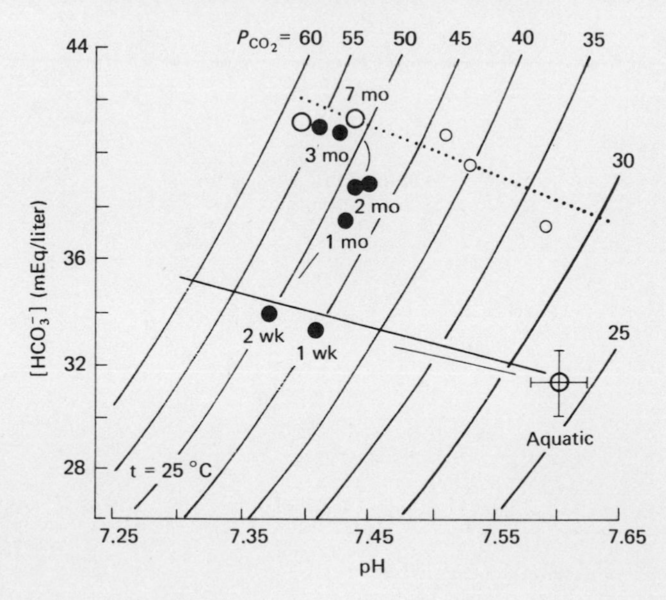

gen ion regulation. Hydrogen ion excretion across the amphibian skin has been demonstrated (Emilio et al. 1970), and, although not directly measured, it probably also occurs across the skin of those fish possessing morphologically similar tissue. Thus gill reduction may involve the relocation of epithelial cells capable of proton pumping from sites in the gill to locations on the general body surface. The relocation would allow the reduction or deletion of the gills and their associated requirement for ventilation but would not eliminate the necessity for water and ionic exchange with the external environment. In amphibians, ionic

Figure 2.22. The gas exchange ratio for the air-breathing organ of a number of air-breathing fish. Closed circles are for animals in water; open circles are for animals in air. *Periopthalmus*, *Trichogaster,* and *Mnierpes* are animals that use their gills or structures derived from gill tissues for aerial exchange. *Hoplerythrinus* has a modified gas bladder for aerial gas exchange.

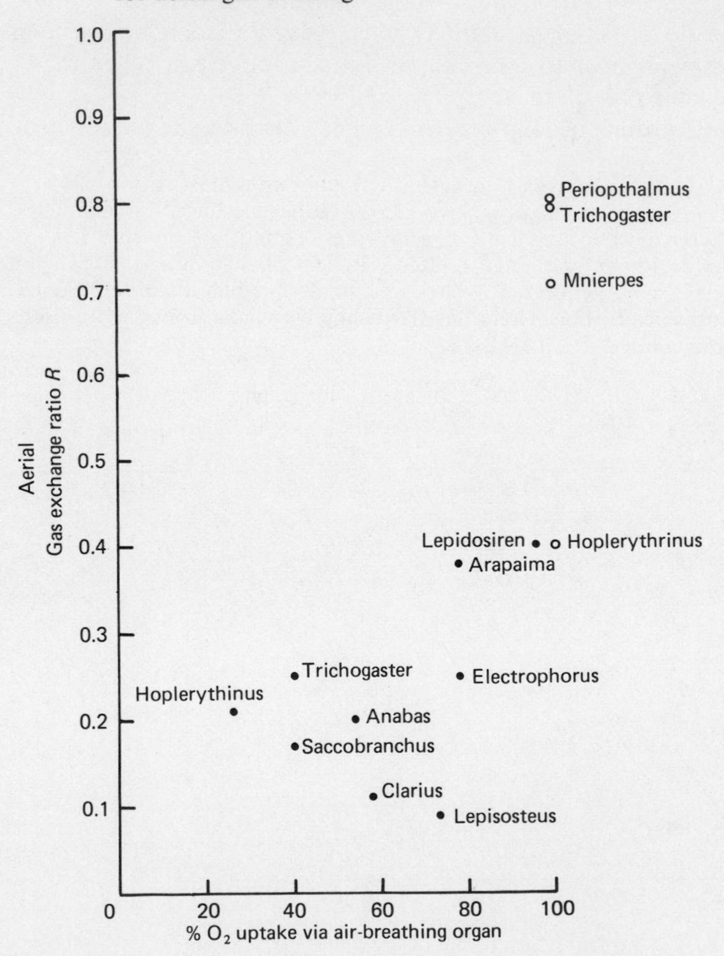

exchange occurs across the ventral surface, with the abdomen and hind limbs representing major sites of salt and water uptake. In some of the more terrestrial amphibians, such as *Bufo,* these ion transporting patches of skin appear to be restricted to areas protected from desiccation by the hind limbs. Skin dehydration will therefore probably reduce the capacity of the animal to regulate pH. This is consistent with the observation that blood pH compensation during hypercapnia was much faster in hydrated than in dehydrated toads (Boutilier et al. 1979a,b). Nearly all modern adult amphibians have freed themselves from the requirement of gills by using lungs for oxygen uptake and moist skin to achieve CO_2 excretion and to control blood pH; however, amphibians

Figure 2.23. Blood pH (pH_B) and carbon dioxide partial pressure (P_{CO_2}) in toads (*Bufo marinus*), snapping turtles (*Chelydra serpentina*), and bullfrogs (*Rana catesbeiana*). Open circles are in vivo data, and filled circles in vitro measurements. Curves are binary buffer model solutions. (Reprinted from Reeves and Rahn 1979, p. 242, by courtesy of Marcel Dekker, Inc.)

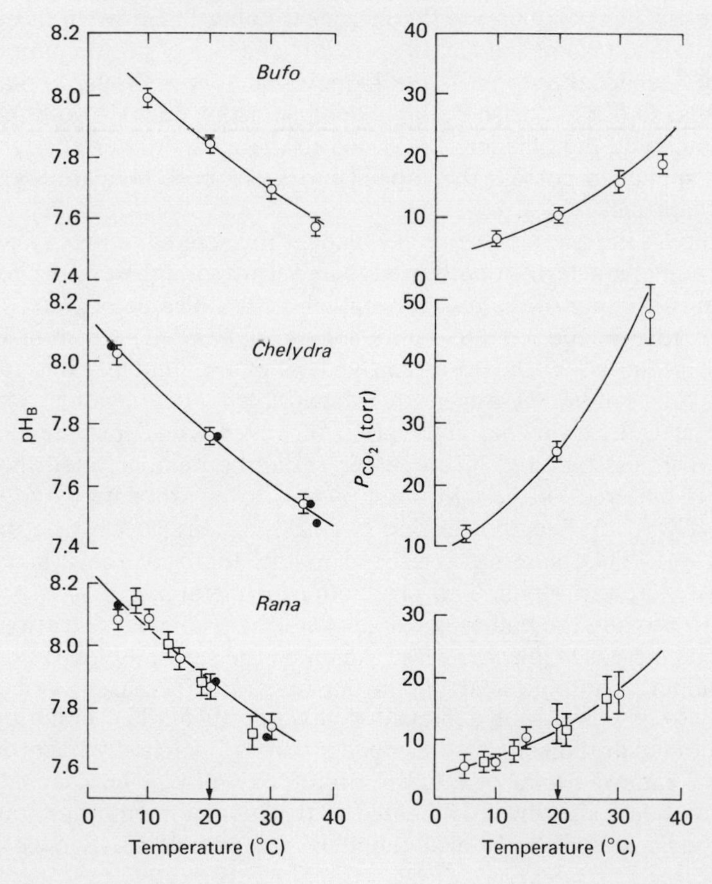

have not completely freed themselves from an aquatic mode of existence.

The relocation of the acid–base regulation site from gills to skin results in important changes in the circulation. In fish all the cardiac output passes through the gills, which are generally in series with the systemic circulation and any air-breathing organ. The cutaneous circulation is in parallel with the lung circulation in amphibians, and both are in parallel with the systemic circulation. The skin circulation can be modulated without necessarily altering systemic or lung blood flow, whereas the gill circulation cannot (see Chapter 3).

The evolution of kidney and bladder function in acid–base regulation

The possible role of the kidney in acid–base regulation in fish is usually not considered, or, when so considered, is given a very minor role (Cameron and Randall 1972; Janssen and Randall 1975; Cameron and Wood 1978). However, it has recently been demonstrated that rainbow trout in freshwater will increase the urinary output of acid when faced with an acid load (Wood and Caldwell 1978). The site of proton pumping has not been localized in fish, but kidney structure is similar to that of Amphibia; so fish may also be capable of excreting acid via both the kidney tubule and the bladder in addition to the gills. All vertebrates, therefore, appear to possess the capacity for acid–base regulation via renal mechanisms.

The ability of the amphibian urinary bladder to secrete hydrogen ions has been well characterized both in vivo and in vitro (Schwartz 1976). At the luminal border of the bladder epithelium a molecule of water is dissociated to produce a proton plus a hydroxyl ion. The proton is excreted with the associated inward movement of a sodium ion. It is not certain if this sodium–hydrogen translocation is an obligatory exchange. The OH^- from the dissociation of water would elevate the cellular pH unless buffered or excreted. It can be demonstrated that this OH^- is buffered by the hydration of CO_2 to bicarbonate ($CO_2 + OH^- \rightarrow HCO_3^-$), the bicarbonate ion having little direct effect on intracellular pH. This buffering is crucial to the ability of the urinary bladder to excrete protons. The production of metabolic CO_2 is not sufficient to provide the buffering capacity during peak acid extrusion, and if CO_2 is increased in a stepwise fashion on the serosal (blood) side of the bladder, titratable acid on the luminal side increases until a maximal rate of pumping is achieved (Schwartz 1976). The buffering action via the hydration of CO_2 is dependent on carbonic anhydrase, as inhibition (Diamox) produces a fall in proton excretion. The Diamox-sensitive carbonic anhydrase is located in the luminal membrane, but some carbonic anhydrase is also found in the cytoplasm (Schwartz

1976). When proton pumping in the toad bladder is inhibited with Diamox, sodium influx also falls. Interestingly, in the trout gill, carbonic anhydrase is also found associated with the water side of the membrane, and Diamox inhibits sodium influx in this tissue as well (Kerstetter and Kirschner 1972). The intracellular bicarbonate formed from the hydration of CO_2 is excreted in the ionic form in the urinary bladder. The bulk of the HCO_3^- moves through the serosal membrane into the blood, although some bicarbonate leaves via the luminal membrane as well. The movement of HCO_3^- through the luminal membrane is dependent on external chloride and appears to be a tightly coupled 1 : 1 exchange. Amiloride, which inhibits sodium uptake, blocks hydrogen ion excretion in conjunction with Na^+ uptake at the gill in trout (Kirschner et al. 1973). The same is also true for the amphibian urinary bladder.

That dehydrated toads can regulate pH, if only poorly compared to hydrated toads, probably represents the ability of their large urinary bladders to remove and store hydrogen ions. Up to one-fifth of their body weight may be as water in the bladder. Amphibian kidney tubules can also excrete acid into the urine, and the combined tubular and bladder acid-secreting mechanism allows *Bufo marinus* to produce urine with a pH of less than 6 during dehydration (Boutilier et al. 1979a,b). Whether or not the kidney actually is utilized to help achieve acid–base regulation is probably dictated by the inherent mechanisms (and/or limitations) of the kidney, and alternate mechanisms available to the organisms. These alternate mechanisms may be modulated by environmental conditions. For example, whereas proton pumping may occur in the very dilute urine of a freshwater fish, it may not be possible in the marine environment where dehydration is a problem. These mechanisms probably evolved under conditions with very small hydrogen ion gradients, such that the ability to secrete protons may be severly limited in the absence of large water fluxes. Thus fish and amphibian kidneys may be limited by their capacity to secrete protons in a concentrated urine, and the capacity to regulate body pH via the kidney may be largely determined by the turnover rate of water. Acid regulation via the skin and kidney requires an immediate supply of water and is an important factor in chaining amphibians to an aquatic existence.

In amphibians the relocation of proton pumping to skin sites may be energetically advantageous owing to the decreased hydrogen ion gradients existing between skin and blood. However, problems of maintaining a moist skin must severely limit the skin as a mechanism to rectify acid–base disturbances in those animals moving away from water. The large bladder volume of the toad probably provides a "water insurance policy" during periods of dehydration and, consequently, may provide a short-term mechanism for storage of protons, which can ultimately be excreted when water is again available (Boutilier et al. 1979b).

Acid–base regulation in mammals is achieved usually via ventilatory adjustments in arterial P_{CO_2} as previously described; however, plasma bicarbonate and hence pH adjustments can be modulated by the kidney. This regulation is a fairly slow process, for example, taking approximately 3–5 days during hypoxic alkalosis in man. Plasma bicarbonate regulation occurs within the kidney tubules by an active process (Burg and Green 1977). The hydrogen ion secretion is similar in some respects to proton pumping in the amphibian and turtle urinary bladder (Sachs 1977), with both processes requiring energy and utilizing some type of ionic pathway through the luminal membrane. Some differences between vertebrate kidneys (i.e., of fish and amphibians and mammals) do exist. In the mammalian kidney, proton pumping is achieved in the tubules without any involvement of the bladder. Presumably the vertebrate kidney, as exemplified by mammals, is better able to regulate salts and acid–base balance at the level of the kidney tubule rather than secondary changes to urine occurring within the bladder. These modifications, along with decreased water loss through the body surface, would obviate the requirement of a large water store, such as the bladder in amphibians.

Although much of our attention to the evolution of the vertebrate kidney has thus far dealt with acid–base regulation, it should also be remembered that the burden of excreting nitrogenous wastes in air breathers is relegated to this organ as well. Over 90% of the ammonia excreted in aquatic fish occurs across the gill (Smith 1929). Even in freshwater, where kidney filtration rates may be quite high, ammonia is excreted across the gill. Unless ammonia is in very low concentrations, it becomes toxic; thus during the evolution of air breathing (and associated problems of water conservation) and terrestrial invasion, animals have resorted to other mechanisms of detoxification. Consequently, either urea or uric acid, which can be accumulated and excreted in the kidney, forms the principal nitrogenous waste product in mammals and birds. It appears that all vertebrates may possess the enzymatic machinery to produce urea; for example the mudskipper produces urea during air exposure (Gregory 1977). However, the exact mechanisms and control of this function in fish are not yet clear.

Acid–base regulation via changes in lung ventilation

In birds and mammals, and probably reptiles, ventilation of the lung is used to regulate the rate of CO_2 excretion and therefore body $[H^+]$ levels. The excretion of CO_2 via the lung requires evolutionary changes in the pattern of CO_2 movement within the blood from the aquatic pattern; in particular, the blood cell must be freely permeable to HCO_3^-. This system of regulating CO_2 excretion does not interfere with oxygen transfer in the oxygen-rich atmosphere, but adds a rapid and powerful mechanism for hydrogen ion regulation, thereby reducing the H^+ load on the kidney and skin. In fact, the kidney alone is able to handle any additional H^+ load, and a moist skin or gills are no longer required for acid–base regulation. Thus the appearance of a mechanism coupling lung ventilation to acid–base regulation lessens the acid–base load on the gill–skin–kidney system that requires fish and amphibians to remain close to water, and paves the way for vertebrates that are much less dependent on water, namely, reptiles, birds, and mammals.

Changes in ventilation to alter CO_2, and therefore pH, require high levels of CO_2 in the blood such that, during an acid load, CO_2 can be reduced by increases in ventilation. Thus the evolution of mechanisms coupling ventilation to acid–base regulation requires not only changes in erythrocytic bicarbonate permeability but also high body CO_2 levels. It is interesting, therefore, to speculate as to whether the high CO_2 levels appeared before or after the increase in erythrocytic bicarbonate permeability. We think they preceded changes in the permeability of red blood cells. For instance, high CO_2 levels are common in many air-breathing fish because of the reduced diffusing capacity of the gills or skin. *Arapaima* has smaller gills per unit weight than *Hoplerythrinus* and has a higher arterial blood CO_2 content and P_{CO_2}. The elevated blood CO_2 levels result in a larger proportion of CO_2 being excreted via the air bladder; however, most of the CO_2 is still excreted via the gills (Figure 2.24) even though arterial P_{CO_2} is 28–30 mmHg in *Arapaima* (Randall et

Figure 2.24. Schematic representation of the pathways of oxygen uptake and carbon dioxide excretion in pirarucu (*Arapaima gigas*). The possible state of the CO_2–bicarbonate system within the plasma is indicated: $CO_2 \rightarrow HCO_3^-$ indicates CO_2 is being hydrated to HCO_3^- at the uncatalyzed rate; $CO_2 \rightleftharpoons HCO_3^-$ infers the CO_2–bicarbonate system is in equlibrium (from Randall et al. 1978b). (Reproduced by permission of the National Research Council of Canada from the *Canadian Journal of Zoology* 56: 977–82, 1978.)

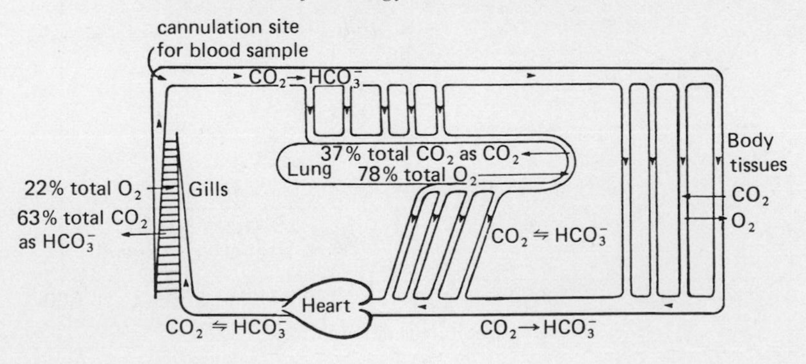

al. 1978b). There is rapid equilibration between blood entering the bladder and bladder gas P_{CO_2} in *Arapaima* (Figure 2.25), which breathes every 3 or 4 min. Bladder gas P_{O_2} seldom falls to oxygen levels recorded in inflowing blood, but P_{CO_2} levels are similar in about a minute after a breath (Randall et al. 1978b). These animals, in which the air-breathing organ is not the major pathway for CO_2 excretion, still respond to increasing CO_2 levels by altering ventilation of the air-breathing organ. *Arapaima* increases breathing frequency in response to injection of CO_2 into the air bladder (Farrell and Randall 1978). The toad, *Bufo,* responds to increases in ambient CO_2 by increasing lung ventilation (Macintyre and Toews 1976). The importance of ventilation in regulating CO_2 levels and therefore pH in bimodally breathing vertebrates, however, is difficult to assess. Body CO_2 levels are adjusted to regulate pH, and ventilation does change in response to CO_2, but this pathway may be of only minor significance as the skin and/or gills are the major pathway for CO_2 excretion. For example, hyperoxia in

Figure 2.25. Changes in P_{O_2} and P_{CO_2} in the lung gas of pirarucu (*Arapaima gigas*) following an air breath. The lines were fitted by eye. The lung *RQ* was calculated from the slope of the lines at different time periods after a breath corrected for lung volume changes after a breath (inserts) (from Randall et al. 1978b). (Reproduced by permission of the National Research Council of Canada from the *Canadian Journal of Zoology* 56: 977–82, 1978.)

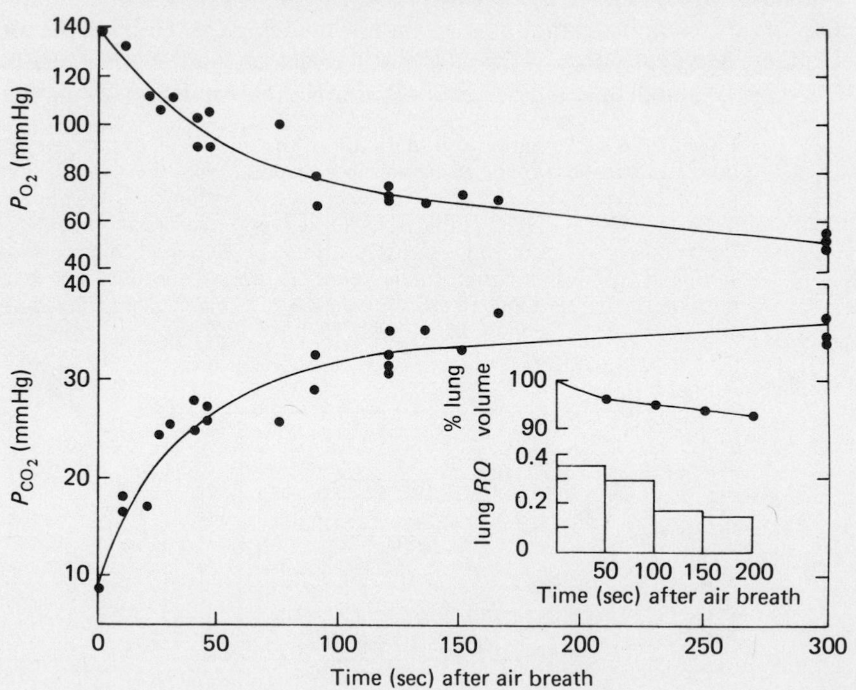

the toad, *Bufo,* causes a marked reduction in lung ventilation, but has no effect on blood CO_2 or pH (Janes 1979).

Thus high body CO_2 levels in vertebrates probably occurred as a result of the evolution of a reduced diffusing capacity of the gills. This then increases CO_2 excretion via the air-breathing organ, and the body CO_2 content can be modulated by changes in ventilation. The major pathway for CO_2 excretion is still the gills–skin; so the capacity of the animals to adjust CO_2 and therefore pH by changes in ventilation is limited. The evolution of a bicarbonate-permeable red blood cell would increase carbon dioxide excretion via the lung and increase the capacity of the animal to regulate pH by changes in ventilation. It would also lead to the uncoupling of CO_2 excretion and ion exchange across the gill–skin system. This presumably presaged the eventual loss of the skin in acid–base regulation. Clearly the selective forces determining the extent of carbamino formation on hemoglobin are also changed.

In conclusion, we think the evolution of mechanisms for adjusting pH by changing lung ventilation was a gradual process involving several steps, namely: (1) a rise in body CO_2 levels associated with a reduction in gill diffusing capacity, (2) the evolution of control mechanisms for regulating CO_2 excretion and therefore pH via lung ventilation, and finally (3) an increase in bicarbonate permeability of the erythrocytes as well as hemoglobin carbamino formation to increase lung CO_2 excretion and therefore the capacity to regulate pH via changes in lung ventilation.

All of these adaptive changes in gas exchange and ion regulation can occur at no greater rate than that at which the cardiovascular and respiratory system evolved to accommodate them. The next two chapters examine these processes of perfusion and ventilation as related to the evolution of air breathing.

3 *Ventilation and perfusion relationships*

Significance of ventilation : perfusion matching

The exchange of respiratory gases between the environment and the metabolic machinery of the cell requires gas transfer between two major convection systems. The respiratory medium (water or air) must be brought into close contact with the respiratory membranes, which represent the first gas diffusion barrier between environment and cell. The convection systems that have evolved to achieve the ventilation of the respiratory surfaces of vertebrates are varied, and reflect the constraints and demands of the respiratory media and the breathing organs, themselves (Chapter 4). Once oxygen has diffused across the respiratory membranes, the second major convection system – the circulatory system – serves to convey this oxygen to the tissues.

In order to facilitate the mass transfer of respiratory gases across these diffusion barriers and between the ventilatory and circulatory convection systems, matching must exist between the delivery of a gas to one side of the membrane and its removal at the other side, having once diffused across the respiratory epithelium. Thus in every vertebrate there is often precise regulation of the ratio between blood perfusion ($\dot{V}_b$) and ventilation with a respiratory medium ($\dot{V}_m$), either water ($\dot{V}_w$) or air ($\dot{V}_{air}$). Changes in this so-called ventilation : perfusion ratio, $\dot{V}_m/\dot{V}_b$, have a profound influence on the ability of a gas exchange organ to function effectively in oxygen transfer. For example, consider briefly oxygen transfer under normal and two extreme conditions of perfusion and ventilation in the simple gas exchange unit depicted in Figure 3.1. If the rates of ventilation and perfusion of this unit are such that the amount of oxygen available to diffuse across the respiratory membrane from one side is exactly matched by the ability of the blood perfusing the opposite side of the membrane to carry this oxygen to the tissues (condition 1), then oxygen transfer relative to its metabolic cost, that of perfusion and ventilation, will be minimized. If, on the other hand, the respiratory unit should suffer an over- or underperfusion relative to ventilation, then the metabolic cost of that oxygen extraction will not be optimal, and the absolute quantity of oxygen removed from the gas exchange unit may well be lower than that occurring with the most favorable ratio of ventilation to perfusion. These arguments apply equally well to gas exchange at the level of the functional exchange unit or at the whole organ level, and are valid for carbon dioxide exchange as well.

There is no "correct" value for $\dot{V}_m/\dot{V}_b$ that can (or should) be found repeatedly in all designs of respiratory gas exchangers. For example, the normal ratio between mean ventilation and mean perfusion at the whole-organ level is approximately 1 : 1 in mammals (West 1977), 2 : 1–5 : 1 in chelonian reptiles (Burggren et al. 1977), and 9 : 1–35 : 1 in aquatic fish (Randall 1970a). The optimal ratio of ventilation volume to perfusion volume is largely determined by the gas capacitances of both the respiratory medium and the blood. In spite of differences in these properties between vertebrate species, there has invariably evolved a relationship between ventilation and perfusion levels such that delivery of a given amount of gas to the respiratory membranes by one convective pump will be matched to the ability of the other convective pump to remove it. Although numerous and often profound changes in the physiological processes of gas exchange have accompanied the evolution and adaptation of air breathing in vertebrates, this matching of gas delivery to transport potential across the respiratory membranes appears fundamental to the efficiency of breathing.

Figure 3.1. Changes in the efficiency of respiratory gas transfer (transfer achieved ÷ energy expended) in three hypothetical gas exchange units, as influenced by variation in the ventilation : perfusion ratio. Condition 1 illustrates optimal gas exchange, with both adequate ventilation and adequate perfusion of the respiratory membrane. In condition 2 (perfusion greatly exceeding ventilation) and condition 3 (ventilation greatly exceeding perfusion) the efficiency of gas transfer falls.

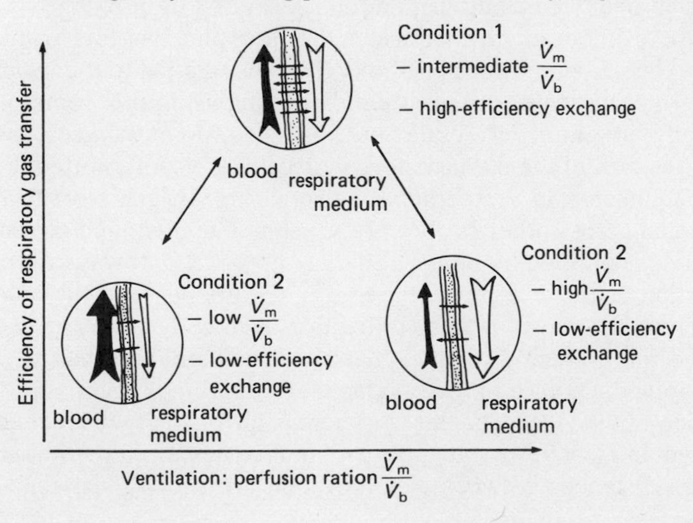

Intermittent breathing and oxygen stores

Most vertebrates must maintain a continual supply of oxygen to at least some tissues. The cerebral cortex, for example, is very sensitive to interruptions in oxygen delivery and carbon dioxide removal. Most unimodally breathing vertebrates, such as strictly aquatic breathers (i.e., most fish), and strictly air-breathing vertebrates, (i.e., birds and mammals), constantly ventilate and constantly perfuse their gills or lungs to maintain a continual supply of oxygen. Bimodally breathing vertebrates, however, show marked variation in air breathing frequency, and some may not breathe for periods of up to 1 or 2 hr. These intermittently breathing animals survive because either large oxygen stores or alternate exchange sites are utilized between breaths.

The oxygen stores in blood and tissues in strictly aquatic fish are small and would last for only a few minutes in the absence of any replacement, even though oxygen consumption rates are low. These animals must constantly ventilate their gills. The evolution of a gas-containing respiratory organ, which may constitute only 5% body volume, immediately increases body oxygen stores by an order of magnitude because it contains oxygen-rich air. An oxygen storage function during interbreath intervals has been suggested for fish gas bladders (Berg and Steen 1965) and the lungs of anuran amphibians (Bastert 1929; Foxen 1964; Jones 1967; Shelton 1970a). Thus, as long as oxygen requirements remain low, air breathing can be intermittent. This is the case for many air-breathing fish, amphibians, and reptiles. Birds and mammals have oxygen requirements that are nearly an order of magnitude greater than those of fish, and thus almost all must breathe continuously even though they have large oxygen stores. Intermittent breathing, defined as lung ventilations separated by prolonged periods when there is no air flow between the lung and the environment, is related, therefore, to the ratio of oxygen consumption to the magnitude of the oxygen store. Intermittent breathing is found commonly in bimodally breathing vertebrates because the ratio of oxygen consumption to the size of the oxygen store is usually low. Intermittent breathing is augmented in vertebrates by increasing oxygen stores; obtaining oxygen across other surfaces (e.g., skin and gills); and/or reducing oxygen consumption by either utilizing anaerobic pathways for energy production or reducing the total energy requirements of the animal.

Different patterns of oxygen utilization from body stores have developed as intermittent breathing patterns and aerial exchange organs have evolved. Oxygen extraction from air-breathing organs is probably perfusion-limited (see Chapter 2). There is little direct evidence for this statement in most vertebrate groups, but it is supported by the fact that diffusion distances between air and blood are usually less than a micron, and oxygen partial pressures in blood leaving the air-breathing organ appear equilibrated with lung gases (West 1977). Thus blood flow determines the rate of oxygen removal from the air-breathing organ between breaths. In many intermittently breathing animals oxygen is rapidly transferred from lung to blood during the first few minutes after a breath. During this time lung blood flow is high, but it then gradually falls, and oxygen uptake from the lung is reduced as the interbreath interval proceeds. In these animals there appears to be a rapid transfer of oxygen from the gas store to blood within the confines of a perfusion-limited system, rather than a slow and steady "metering out" of aerial oxygen. This rapid transfer of oxygen from the bladder or lung to blood is typical of animals in which gills or skin is retained as a secondary source of oxygen uptake.

Many reptiles, whose abilities to exchange respiratory gases at extra-pulmonary sites are in doubt, also ventilate their lungs in a periodic fashion (Figures 3.2 and 3.3). Figure 3.2 illustrates lung and arterial blood oxygen partial pressures during a prolonged dive in the turtle *Chrysemys scripta*. The initial third of the dive was characterized by a rapid fall in arterial P_{O_2} but little or no change in lung P_{O_2}, indicating that the lungs were not being perfused, and the blood oxygen store was being depleted. After $\frac{1}{2}$–1 hour of diving, however, there was a precipitous fall in lung P_{O_2} and a large increase in arterial P_{O_2}, indicating that the pulmonary vascular bed was now perfused, and oxygen was transferred from the lung to the blood. Thus it appears that during a dive pulmo-

Figure 3.2. Changes in the partial pressure of O_2 in lung gas (P_{AO_2}) and femoral artery (P_{aO_2}) during an extended voluntary dive in the turtle *Chrysemys scripta*. This particular pattern is evident in approximately 20% of all dives by this species. Shaded areas represent periods of lung ventilation. See text for further explanation (from Burggren and Shelton 1979).

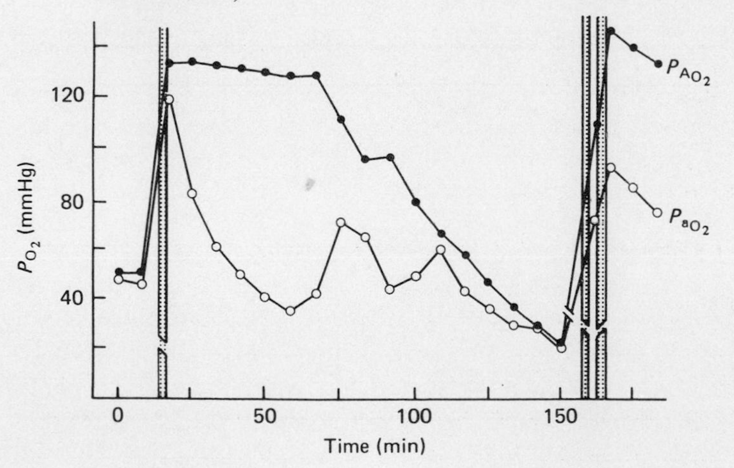

nary perfusion can be stepped up to transfer oxygen from the lung to the blood, and then greatly reduced or stopped to retain oxygen in the lung while the blood oxygen stores are being depleted. Although some of the partial pressure changes recorded in these turtles could have been accounted for by changes in intracardiac blood shunts, the magnitude of the P_{O_2} changes inevitably leads to the conclusion that the lung of *Chrysemys* can function as an oxygen store, to be tapped periodically during a dive.

Among a wide range of fish, amphibian, and reptilian bimodal breathers, there is a clear tendency for animals with an increasing dependency upon aerial respiration to retain larger oxygen reserves in their aerial exchange organs even at the end of a period of apnea (Figure 3.4). Air breathing, then, especially in more terrestrial forms, often recommences at a point in time when two-thirds or more of the oxygen initially in the aerial exchange organ at the onset of apnea remains to be taken up (Figure 3.5). The limited data available indicate that the oxygen partial pressures in the blood perfusing the gas bladder or lung(s) are usually sufficiently low during interbreath intervals to allow for a continued oxygen extraction, but it is the rate of perfusion of these

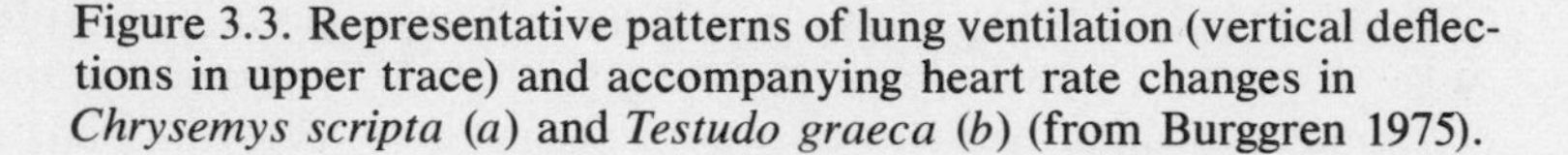

Figure 3.3. Representative patterns of lung ventilation (vertical deflections in upper trace) and accompanying heart rate changes in *Chrysemys scripta* (*a*) and *Testudo graeca* (*b*) (from Burggren 1975).

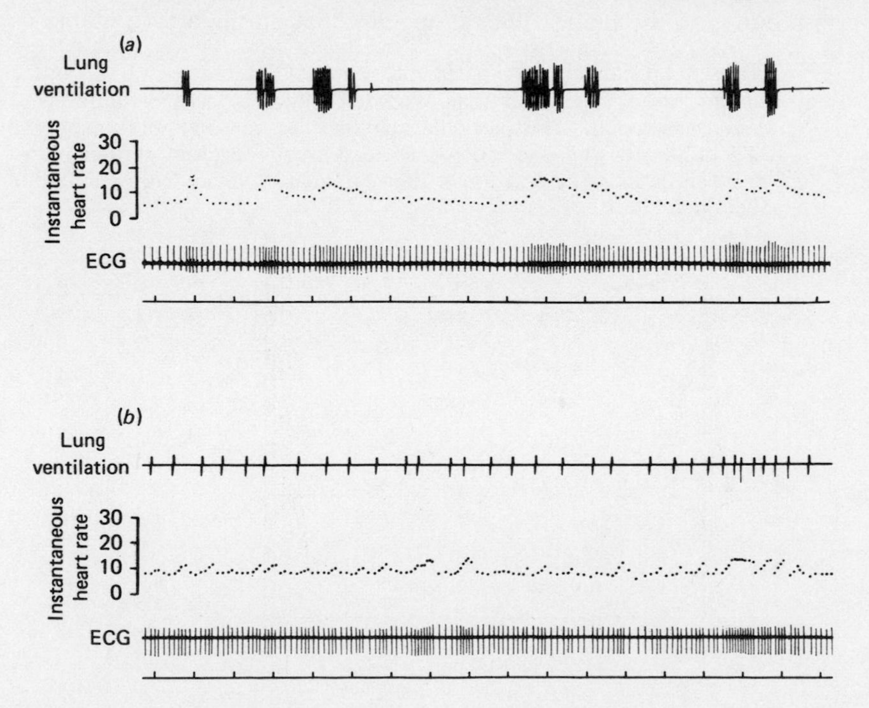

Figure 3.4. The relationship between the percentage of oxygen uptake that results from air breathing and the depletion of oxygen from the aerial exchange organ during apnea in various bimodally breathing fish, amphibians, and reptiles.

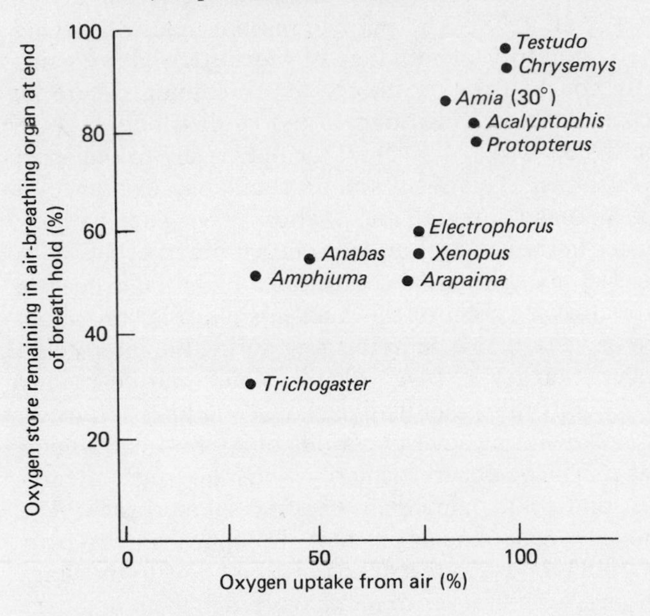

Figure 3.5. Relationship between apnea length and the rate of oxygen depletion from the aerial exchange organs of some intermittently air-breathing fish and reptiles. The point on each line indicates the average time when apnea is terminated by an air breath.

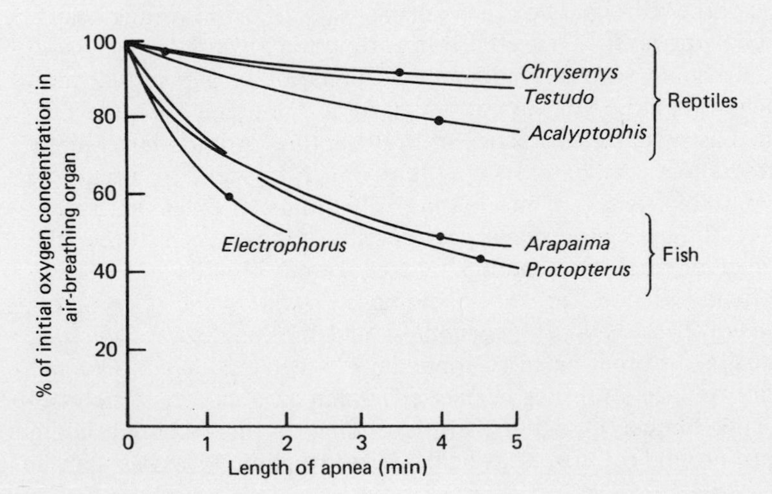

organs during apnea that becomes the major limiting factor to gas exchange. This is clearly evident in the intermittently breathing turtle *Pelomedusa subrufa,* which has a pulmonary artery P_{O_2} of 30–35 mmHg and an end tidal (alveolar) P_{O_2} of 120–140 mmHg, irrespective of diving or breathing (Glass et al. 1978). The partial-pressure gradient driving oxygen across the respiratory membranes of the turtle thus remains large and relatively constant during intermittent breathing. There is, however, a fine control over cardiac output and its distribution, such that pulmonary perfusion is very transient, limited largely to the short periods of lung ventilation. This results in an oscillating exchange of gases in the chelonian lung (Burggren and Shelton 1979). The changing nature of gas transfer in the lungs of an intermittent breather has been directly demonstrated by Emilio and Shelton (1974) investigating *Xenopus*. The respiratory quotient of this anuran's lungs is only about 0.2; so any respiratory gas exchange in the lung during the interbreath interval will produce changes in lung volume, which can be readily measured. Figure 3.6 illustrates cumulative changes in lung volume in an unrestrained *Xenopus* during several periods of intermittent breathing. Pulmonary gas exchange occurs primarily when the frog is breathing at the surface, and for a minute or so after submergence. The absence of pulmonary gas exchange later in apnea in *Xenopus* doubtlessly stems from the severe restriction of blood flow to the lungs under these conditions, which serves to retain oxygen in the pulmonary store for later use.

This pattern of oxygen extraction, in which blood flow restrictions to the gas exchanger retain oxygen in the lung or gas bladder, relates not so much to the physiology of aerial respiration as perhaps to behavioral and environmental factors. It is very unlikely that an animal highly dependent on its lung oxygen store during apnea would allow a large depletion of the pulmonary oxygen during a dive of only average length, for the following reason. If, as the animal returned toward the surface to breathe, it should suddenly encounter a predator or a potential food item, then it might be very important to forgo breathing for other purposes such as predator avoidance or prey capture. Additionally, many air breathers (e.g., anabantids) may temporarily suspend air breathing during courtship behavior and mating. This unusual extension to the normal breath-holding interval is only possible if the oxygen store in the lung or bladder has been kept high by frequent air breaths and/or large tidal volume. Hence, in the chelonians *Testudo* and *Chrysemys* (= *Pseudemys*), the snake *Acalyptophis,* and the lungfish *Protopterus,* the lung oxygen store is usually "topped off" when less than 20% of the pulmonary oxygen store has been used, much as a motorist might fill his fuel tank frequently rather than depending on there being a filling station just around the corner as he begins to run low. Whereas normal interbreath intervals in the above animals using mainly aerial respiration are thus relatively short (Figure 3.3), the potential length of the interbreath interval is much longer.

It has been suggested (Romer 1972) that the gas bladder of fish arose first for purposes of respiration, and has only been secondarily modified for buoyancy control in modern water breathers. We believe that the buoyancy and respiratory functions could have been served simultaneously with a single bladderlike structure in the ancestral air-breathing fish, and that selection pressures then operated to separate these functions in certain extant air breathers and other aquatic forms. The incorporation of a volume of gas into the body of a fish will not only enhance oxygen stores but also influence buoyancy. If the air-breathing organ is used largely for O_2 uptake, then, as O_2 is removed and only a little CO_2 added to the gas phase, the fish becomes less buoyant and sinks (Gee and Graham 1978). The animal must rise in the water column to breathe, and O_2 utilization and hydrostatic considerations inter-

Figure 3.6. Aerial oxygen consumption (circles) and carbon dioxide production (squares) in *Xenopus* (85 g) as measured by analysis of a gas bubble from which the animal breathed. The alternately continuous and broken line plots the measured fall in volume of the lungs and gas bubble as it occurred during each breathing–diving cycle. The upper part of the figure shows the considerable variation in lung volume at the end of every diving period; the animal can dive with its lungs inflated to different extents (from Emilio and Shelton 1974).

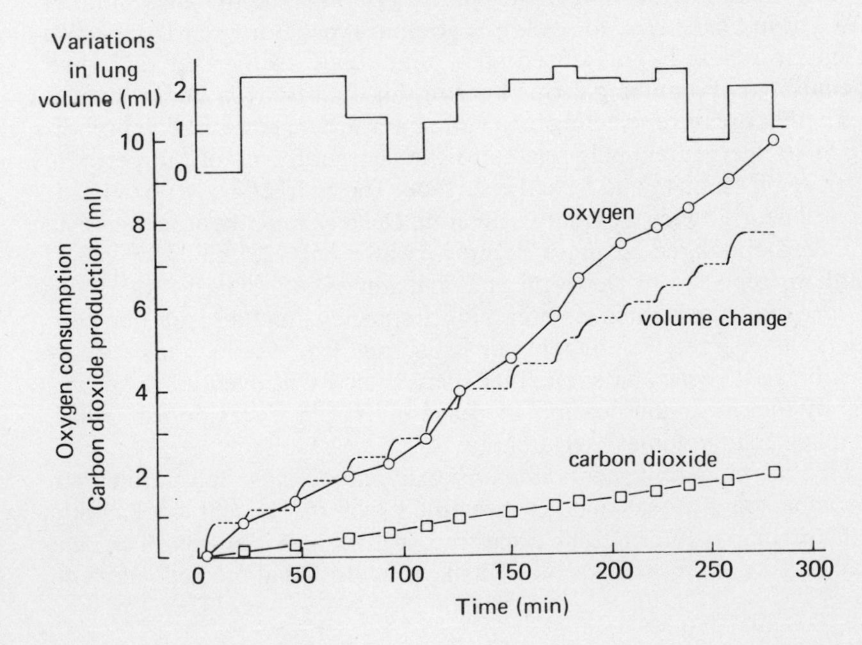

act to determine the extent of O_2 removal and therefore bladder/lung volume prior to a breath, as well as breathing frequency.

Intermittent breathing in bimodally breathing animals seldom affects CO_2 and pH levels in the body because CO_2 is excreted via other pathways in a continous manner, and small oscillations in CO_2 excretions are buffered by the presence of large CO_2 stores in the body. In general, intermittently air-breathing animals have much larger body CO_2 stores than continuously water-breathing vertebrates. Also in many bimodally breathing vertebrates the gills and/or skin are used for CO_2 excretion in water and are continuously ventilated, whereas the air-breathing organ is used for oxygen uptake and is ventilated intermittently (Johansen and Lenfant 1968; McMahon 1969; Farber and Rahn 1970; Singh and Hughes 1973; Randall et al. 1978a,b; Farrell and Randall 1978; Burggren 1979).

Selection pressures for and against the evolution of intermittent breathing

There are several pressures tending to select for longer interbreath intervals, and hence a lower breathing frequency, in air-breathing fish. Surfacing for air breathing carries with it the risk of predation at the air–water interface, which often may be more intense than that under water. It may not be coincidental that the heads of many Amazonian air-breathing fish are heavily armored.

Turtles can tolerate long breath-holding periods, up to a maximum of 1 to 2 hr in *Chrysemys,* for example (Belkin 1964; Burggren 1975), even when air-exposed. This observation may relate to the fact that such animals are often diving forms hunting for long periods under water.

Another pressure selecting for a more intermittent pattern of air breathing in submerged animals relates to the energetic cost of the periodic vertical migration to and from the surface. The energetic costs of surfacing are nearly twice as great in juvenile *Ophicephalus* kept in water 40 cm deep compared to those living in water only 2.5 cm deep (T. J. Cardian, reported in Dehadrai and Tripathi 1976). Clearly, then, any factor tending to reduce air-breathing frequency should result in a considerable energetic saving in air-breathing fish. Reductions in air-breathing frequency in submerged vertebrates will eventually be limited by the maximum possible increase in oxygen stores, in particular, tidal and lung volumes (see Chapter 4).

Air-breathing movements in many fish, amphibians, and reptiles are accompanied by conspicuous movements of the buccal floor, body wall, or limbs. Intermittent breathing, with longer periods of apnea, may be selected for in those "ambush" predators that are dependent on remaining unobtrusive. Of course, similar arguments apply to intermittently breathing animals that are preyed upon.

Selection pressures also operate to reduce apnea length in aquatic air breathers. Increases in breathing frequency will reduce oscillations in arterial blood P_{O_2} and P_{CO_2} and so ultimately in O_2 delivery to the tissues and CO_2 removal from them. This will tend to reduce the necessity for gross cardiovascular readjustments to preserve a metabolic homeostasis.

Different selection pressures must be considered for the retention of intermittent breathing in terrestrial animals. The oxygen demand of amphibians and reptiles is low, and the oxygen storage capacity of the lungs is high; so intermittent breathing remains feasible. Moreover, Milsom and Jones (1979) have suggested that, in reptiles, the work of breathing is clearly optimized over a narrow range of tidal volume and breathing frequency, and the energetic cost of intermittent breathing may well be lower than that of a high-frequency, low-amplitude pattern consistent with continuous breathing. However, lung ventilation in the tortoise *Testudo graeca,* for example, consists of regularly spaced single breaths with interbreath intervals on the average one-tenth those of the turtle *Chrysemys scripta* (Burggren 1975) (Figure 3.3). One obvious difference between species is that lung ventilation in the tortoise is not restricted to short periods of emergence from water as in the turtle. Free from this restraint, this chelonian has evolved a more regular pattern of breathing, as have some of the more terrestrial snakes and lizards. With a comparatively low metabolic rate and large O_2 stores, however, there appears to have been no further selection in these terrestrial reptiles for continuous lung ventilation, as occurs in endotherms.

Respiratory–cardiovascular coupling

The ability of most intermittent breathers to circumvent perfusion of the respiratory gas exchanger during apnea is of great importance in maintaining the efficiency of O_2 and CO_2 transport. Toward the end of prolonged periods of apnea the potential for aerial gas exchange will be reduced as the respiratory partial pressures in gas and venous blood come closer together; at this time the metabolic cost of perfusing the aerial exchange organ with blood may outweigh the benefit of any continued gas exchange that could be achieved. By exerting control over perfusion rate, through a combination of reduced cardiac output and redistribution of blood flow away from the exchanger, a delicate balance between respiratory gas demand and delivery can be maintained. This variable flow of blood to the gas bladder, lungs, or skin is a product of several cardiovascular adjustments, most of which can only occur if the circulation is undivided.

The heart rate response

An increase in the frequency of the cardiac pump, particularly in animals with intrinsically slow heart rates where ventricular filling during diastole is not limiting, is a very effective method of altering cardiac output, and hence of potentially altering perfusion of the respiratory membranes. Many of the extant air-breathing fish tend to show their highest heart rates during active ventilation of the aerial exchange organ (see review by Johansen 1970; Singh and Hughes 1973). In some of these fish, the increase in heart rate is very closely synchronized with the ventilatory event, whereas in others the initial increase and/or final decrease in heart rate may be less closely associated with the onset and termination of ventilatory movements.

Changes in heart rate concomitant with lung ventilation are also evident in nearly all of the intermittently breathing amphibians and reptiles that have been examined (see Shelton 1975; White 1976). Figure 3.3, for example, documents the tachycardia during lung ventilation in an air-exposed aquatic turtle, *Chrysemys scripta,* and a terrestrial tortoise, *Testudo graeca.* Although these two reptiles come from different habitats and exhibit quite different breathing patterns, they have a largely similar pattern of coupling of heart rate and lung ventilation, reflecting the importance of the fundamental cardiac response in matching perfusion to circulation.

Referring generally to the decrease in heart rate in amphibians and reptiles (or in many intermittent breathers, for that matter) during periods of apnea as a "diving bradycardia" is unfortunate; for it shifts the emphasis away from the fact that cardiac events are matched in the strictest sense to ventilatory events, not to behavioral events, of which breath holding during diving is only one of many. Large fluctuations in heart rate during intermittent breathing occur not only in terrestrial, even desert-dwelling, reptiles, but also continue unabated during air exposure even in the most aquatic of air-breathing fish, amphibians, and reptiles. Moreover, active air breathing in undisturbed turtles, for example, occupies only 2–15% of their total activity (Belkin 1964; Burggren 1975; Milsom and Jones 1979), so that means heart rate for an individual animal closely approximates that during apnea (diving) rather than breathing. Hence, the term *ventilation tachycardia* has been adopted to better describe the ventilation-correlated changes in heart rate during intermittent breathing in amphibians and reptiles (Belkin 1964; Gaunt and Gans 1969; Burggren 1975). This term emphasizes that the heart rate evident during lung ventilation in many intermittent breathers is an increase in rate above "normal" levels.

Cardiac output and its redistribution in intermittent breathers

Fish. The long-term distribution of total cardiac output as well as the derivation of blood (i.e., oxygenated or deoxygenated) perfusing the accessory gas organs of air-breathing fish varies considerably, mainly as a function of (1) the degree of specialization of the aerial exchange organ and the vascular system supplying it, (2) the relative importance of this organ in oxygen uptake, and (3) the extent of gill surface area reduction. It is not surprising, therefore, that there have evolved a large number of variations on many cardiovascular themes in intermittent and bimodal breathers.

A review of cardiovascular function in the extant air-breathing fish in terms of cardiac–ventilatory coupling would suffer from the paucity of physiological data on the subject. However the changes in the performance of the cardiovascular system during intermittent breathing are reasonably well understood for at least three quite different fish or groups of fish, *Electrophorus, Hoplerythrinus,* and the Dipnoi.

Electrophorus possesses a highly vascularized buccal cavity that is ventilated every 2–5 min with air. Approximately 80% of this eel's oxygen uptake is through aerial gas exchange, with a similar percentage of carbon dioxide eliminated into the water (Farber and Rahn 1970). Hence, as in other air-breathing fish, the aerial exchange organ functions mainly to acquire oxygen. The gills are reduced in size to minimize oxygen loss from the blood in the secondary lamellae to the water, but they are still retained for carbon dioxide elimination and ion exchange. The circulation to the buccal organ and to the gill arches is effectively located in parallel. The blood draining the buccal cavity passes directly back to the heart for recirculation in the ventral aorta.

Changes in the percent of total cardiac output perfusing the mouth during a single breath-holding period in *Electrophorus* are shown in Figure 3.7. During apnea in *Electrophorus,* buccal gas P_{O_2} falls rapidly, and P_{CO_2} rises. Cardiac output decreases, and the percent perfusing the mouth becomes progressively reduced, but gill perfusion increases relatively. The calculated percent of cardiac output distributed to the buccal cavity during the interbreath interval is greatest when buccal gas P_{O_2} is the highest (Figure 3.8), which is consistent with an effective matching of oxygen delivery by the ventilatory convective pump to the ability of the circulatory convective pump to transport this gas. The net result of these cardiovascular adjustments during intermittent breathing in *Electrophorus* is that dorsal aorta P_{O_2} and P_{CO_2} show only small fluctuations. The oxygen content of blood perfusing the tissues is stable and almost always 70–80% saturated, and thus oxygen delivery to the tissues is determined by blood flow.

The jeju, *Hoplerythrinus,* like *Electrophorus,* exploits air breathing to obtain oxygen. *Hoplerythrinus,* however, has evolved a vascularized gas bladder. A narrow blood vessel connects the dorsal aorta with the two posterior arches, which give rise directly to the coeliac artery, which perfuses the gas bladder (Figure 3.9a). The impedance of this narrow connecting vessel is apparently high, for most of the 20–40% of cardiac output that enters gill arches three and four perfuses the gas bladder, whereas the remaining 60–80% that enters gill arches one and two perfuses the systemic circulation via the dorsal aorta (Smith and Gannon 1978). An increase in heart rate accompanies the onset of air breathing in *Hoplerythrinus* (Figure 3.9b); cardiac output, however, remains unchanged, indicating reciprocal changes in stroke volume and heart rate. Blood flow to the gas bladder almost doubles immediately following an air breath; so blood flow in the dorsal aorta must be reduced during air breathing and the first seconds of apnea (Figure 3.9). Within a few minutes into the interbreath interval, however, blood flow in *Hoplerythrinus* is actively redistributed away from the gas bladder and back to the systemic arterial circulation. The redistribution of a constant cardiac output between the aerial exchanger and the systemic

Figure 3.7. Changes in O_2 and CO_2 tensions of the mouth gas (M), systemic arterial blood (a), and blood from the jugular vein (JV) between two consecutive air breaths in *Electrophorus*. Tabulated above the figure are computed values of the gas exchange ratio (RE), the fraction of the total cardiac output going to the mouth (V_M/V_T), and the fraction to the systemic arteries (V_a/V_T). The vertical arrows indicate the gas partial pressures used for these calculations (from Johansen et al. 1968b).

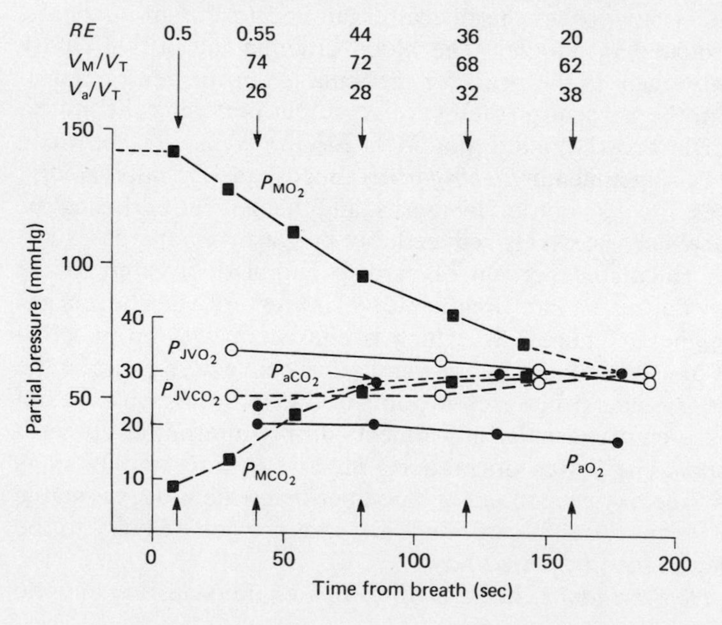

circulation during air breathing could be achieved by general vasoconstriction of the systemic circuit, a vasodilation of the aerial exchange organ circuit, or a selective control over branchial arch perfusion. These events could occur separately or in concert with one another to redistribute cardiac output during intermittent breathing. The specific cardiovascular mechanisms responsible for the redistribution of cardiac output during intermittent air breathing in *Hoplerythrinus, Electrophorus,* and the other air-breathing teleosts have yet to be fully described.

In both *Hoplerythrinus* and *Electrophorus* there is little anatomical provision for the separation of oxygenated and deoxygenated blood in the heart and ventral aorta, but there is a potential in both species for oxygenated blood, leaving the aerial exchange organ, to be circulated via the gills to the body tissues. As in *Electrophorus,* dorsal aorta P_{O_2} in *Hoplerythrinus* fluctuates by only a few millimeters of mercury during intermittent ventilation of the gas bladder. The distribution of cardiac output between the systemic and bladder circulations is controlled presumably to maximize extraction of oxygen from the bladder and to

Figure 3.8. Relationship between the fraction of the cardiac output going to the mouth (V_M/V_T) and the partial pressure of oxygen in the mouth gas (P_{MO_2}). The three curves were obtained while the animal was breathing a hyperoxic gas mixture (O_2), ordinary ambient air (Air), or a nitrogen-enriched, hypoxic gas mixture (N_2). The mouth gas and blood samples underlying the calculations were obtained during the interval between consecutive breaths for each of the three established curves (from Johansen et al. 1968b).

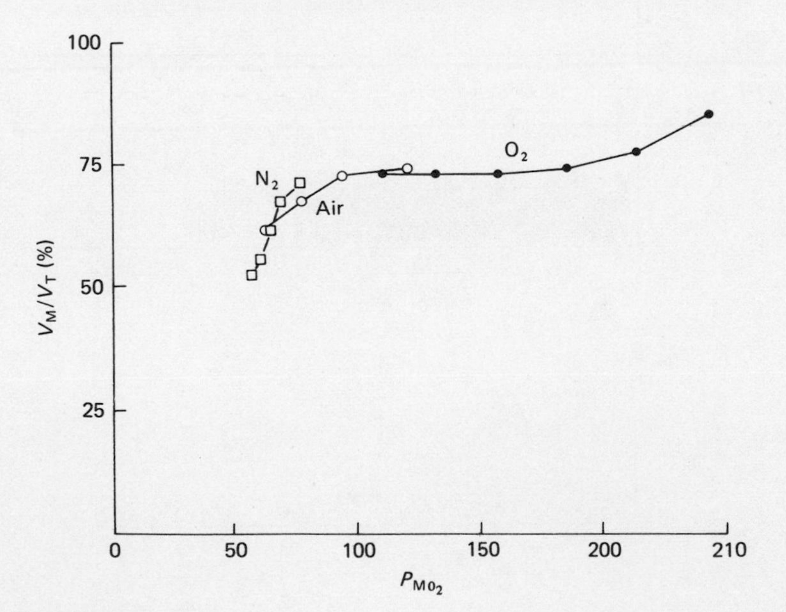

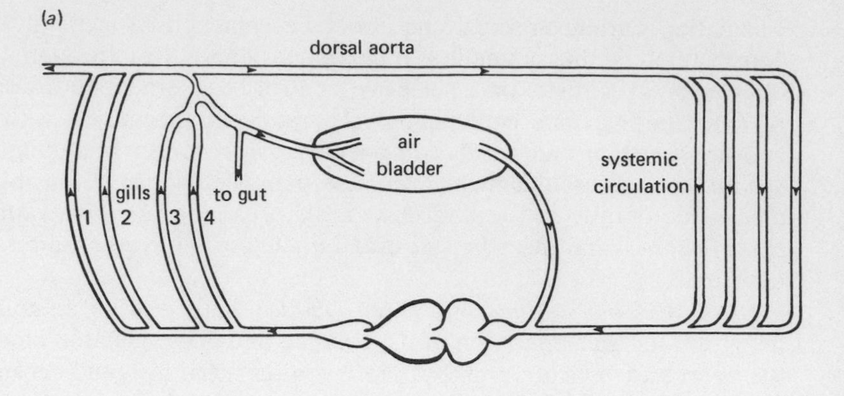
(a)
dorsal aorta
air
bladder
systemic
circulation
gills
to gut
1
2
3
4

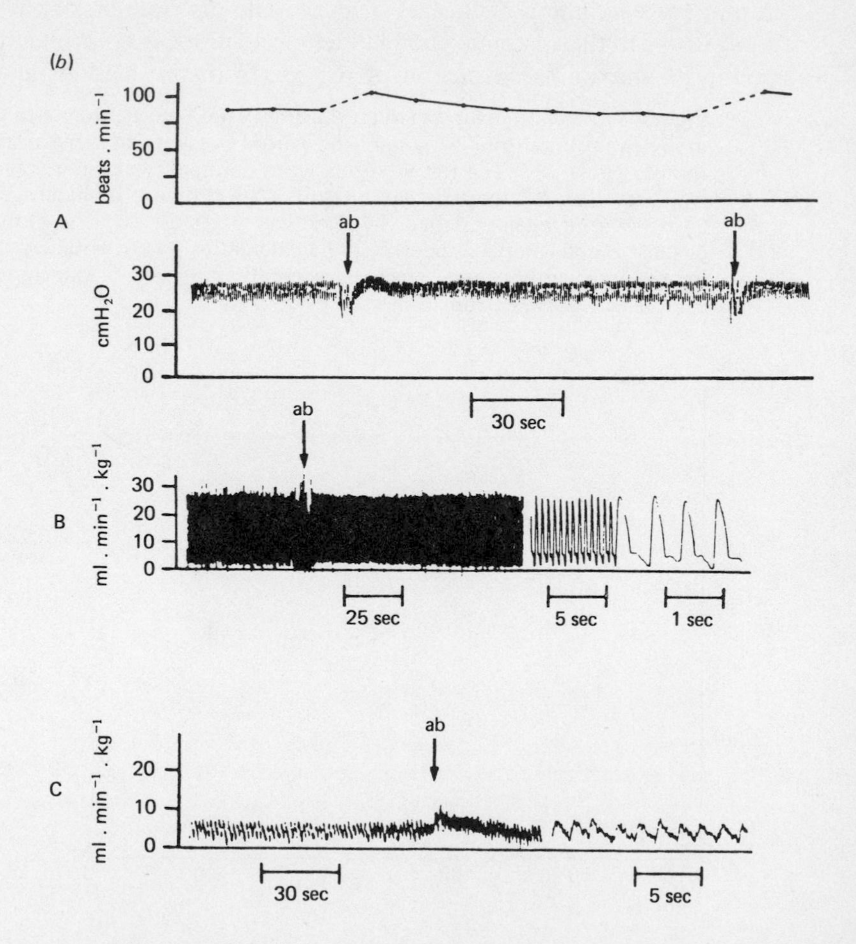
(b)
beats . min⁻¹
100
50
0
A
ab
ab
cmH₂O
30
20
10
0
30 sec
ab
ml . min⁻¹ . kg⁻¹
30
20
10
0
B
25 sec
5 sec
1 sec
ab
ml . min⁻¹ . kg⁻¹
20
10
0
C
30 sec
5 sec

minimize oscillations in oxygen delivery to the tissues. Of course, in those aquatic air breathers that retain either functional gills or a permeable skin, admixture of branchial or cutaneous blood will tend to dampen oscillations in systemic oxygen levels resulting from a decrease in aerial oxygen extraction during apnea.

The Dipnoi, the earliest surviving sarcoptrygian group, have a highly specialized cardiovascular system (Figure 3.10) adapted for bimodal respiration, which resembles the amphibian situation, more so than that of any of the Actinopterygii. As in *Electrophorus* or *Hoplerythrinus,* oxygenated blood draining from the aerial exchange organ of lungfish can either pass into the afferent branchial circulation and then directly to the body tissues, or it can be recirculated to the pulmonary respiratory membranes (Figure 3.10). Distinguishing the lungfish from the teleosts, however, is a discrete pulmonary vein that empties into one side of a partially divided atrium. The lungfish ventricle is also partially

Figure 3.10. Schematic diagram of the circulation of the lungfish *Protopterus*. Arrows indicate blood flow, which will vary in direction and magnitude during intermittent lung ventilation. Note the similarities between the arterial arrangement in *Protopterus* and in *Hoplerythrinus* (Figure 3.9*a*) (after Johansen et al. 1968a).

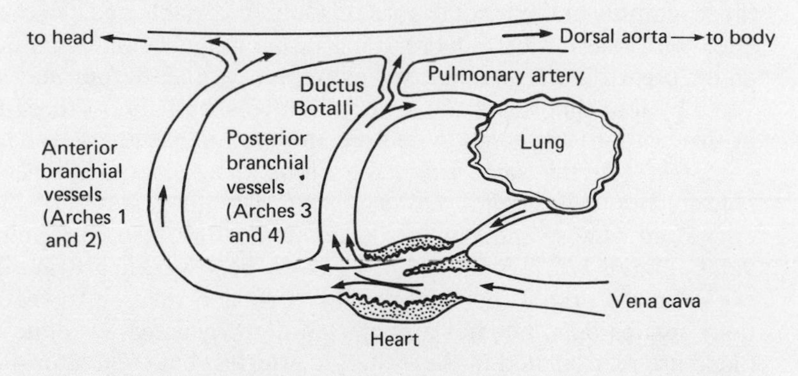

Figure 3.9. (*opposite*) (*a*) Schematic diagram of the circulation of *Hoplerythrinus unitaeniatus*. Gill arches three and four do not drain directly into the dorsal aorta, but via a narrow vessel instead. The air bladder supply is from a branch of the coeliac artery. Venous drainage of the air bladder returns directly to the heart. (*b*) Typical records of cardiovascular responses to air breathing in *H. unitaeniatus* in water. (A) Dorsal aortic pressure (lower) between successive spontaneous air breaths (ab) and accompanying heart rate (upper). (B) Pulsatile ventral aortic flow before and after an air breath (ab). (*c*) Pulsatile blood flow to the air bladder before and after a breath (from Farrell 1978). (Reproduced by permission of the National Research Council of Canada from the *Canadian Journal of Zoology* 56: 953–8, 1978.)

septate and is heavily trabeculated. Additionally there is a complex set of conal valves preceding a very short ventral aorta (see Satchell 1976). These features, probably by establishing and maintaining laminar flow in the central circulation (Johansen and Hol 1968), permit a surprisingly effective separation of pulmonary arterial blood from systemic venous blood.

The extent of this separation among the three dipnoian genera appears related to the degree of aerial respiration. The Australian lungfish *Neoceratodus,* for example, ventilates its lung very infrequently, and retains well-developed gills for oxygen uptake as well as carbon dioxide excretion. Blood flow to the lung in normoxic water is usually less than one-third of cardiac output (Johansen et al. 1968a). Furthermore, there is no tendency for a preferential systemic distribution of blood returning from the lung, and similarly there is no clear distribution of systemic venous blood to the pulmonary artery (Johansen and Hanson 1968). In *Protopterus* and *Lepidosiren,* however, which have reduced gills and frequently ventilate their lung to achieve a large aerial uptake of O_2, there is definite separation of deoxygenated and oxygenated blood to the pulmonary and systemic circuits, respectively (Johansen and Lenfant 1967; Lenfant and Johansen 1968).

The distribution of pulmonary and systemic venous return as well as the magnitude of the cardiac output is highly variable and closely correlated with intermittent air breathing in the lungfish. Immediately after an air breath in *Protopterus,* for example, cardiac output may be elevated by four times over the prebreath levels, and almost three-fourths of this output is directed to the recently ventilated lungs (Johansen et al. 1968a). At the same time, the blood that does pass through the anterior branchial arches and on to the systemic vascular bed may be composed almost entirely of oxygenated blood from the pulmonary veins (Figure 3.11). As apnea progresses, however, heart rate falls, as does cardiac output, and pulmonary perfusion relative to that of the body tissues falls. Greater amounts of deoxygenated systemic venous blood are recirculated in the systemic arteries, but even after 4–5 min, which is the average length of the interbreath interval for *Protopterus* (Lenfant and Johansen 1968), a considerable degree of blood separation in the central circulation is still maintained, for some 65% of anterior branchial flow still consists of pulmonary venous blood.

Variations in the impedance balance between the two circuits as well as between anterior and posterior gill arches will be of great importance in controlling cardiac output distribution in lungfish, just as in *Hoplerythrinus* and *Electrophorus*. For example, vasomotor activity ensuring a lower pulmonary than systemic vascular impedance, particularly during active air breathing, will favor lung perfusion over systemic perfusion in lungfish. However, any postulated mechanism of cardiac output distribution between lungs and trunk must additionally take into account the specific vasomotor responses of the ductus Botalli and the gills, as well as any structural heterogeneity between the anterior and posterior gill arches. The posterior branchial arches of *Protopterus* bear numerous filaments with lamellae, as opposed to the denuded anterior arches, which serve as shunt vessels to the systemic circulation. Although lung ventilation is intermittent, continued perfusion of these posterior arches is important for CO_2 elimination (Lenfant and Johansen 1968) and potentially can occur independent of variations of pulmonary perfusion because the ductus Botalli can serve to shunt this blood directly into the dorsal aorta (Figure 3.10). Recently, Laurent et al. (1978) have demonstrated that acetylcholine constricts the extrinsic pulmonary artery of *Protopterus* but dilates the ductus Botalli. Thus the cholinergic vasomotor responses of the two vessels would act in concert to direct efferent branchial blood from the two posterior arches into the dorsal aorta rather than the lung (Figure 3.10), which would be an appropriate response during apnea.

The Dipnoi apparently were established by the Devonian period. They have since become adapted to fill successfully particular niches, which are probably quite remote from those of the crossopterygians, whence arose the terrestrial tetrapods. Yet, in several general respects, the circulatory anatomy and physiology of the Dipnoi and the present-

Figure 3.11. Time course of the proportion of pulmonary flow to total flow perfusing the anterior gill-less branchial arteries during an interval between air breaths in *Protopterus aethiopicus* (from Johansen et al. 1968a).

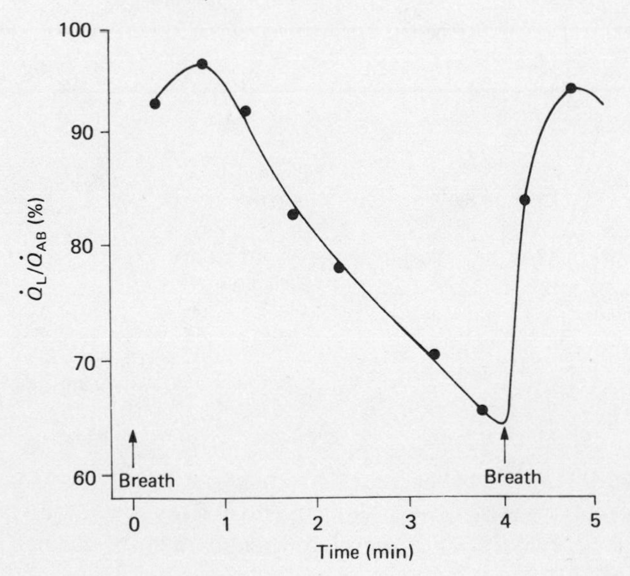

day Amphibia and Reptilia differ only in detail. This similarity suggests that there are many examples of convergent evolution to be found among these animals, and serves to emphasize that certain conditions seem fundamental to bimodal respiration, intermittent breathing, and survival during short- and long-term air exposure. In both the lungfish and the ectothermic tetrapods, for example, a variable proportion of the cardiac output perfuses the gas exchanger, because in both groups the circulation of the systemic vascular bed and that of the gas exchanger are located in parallel, and are perfused by a single cardiac pump. Pulmonary and systemic venous blood return to the amphibian and reptilian heart in completely separate channels, but the ventricles of these vertebrates remain functionally undivided; so some admixture of these streams of blood can occur (Figure 3.12), as in all air-breathing fish.

Figure 3.12. Schematic diagram of the heart and arterial circulation of (*a*) an anuran amphibian and (*b*) a noncrocodilian reptile. The cardiac anatomy, particularly that of the reptile, is not intended to portray actual intraventricular structures, but simply to show gross anatomical divisions. Arrows indicate regions of the circulation where active vasomotor activity and electrical and mechanical cardiac events act to influence the relative balance between blood flow to the systemic tissues and to the gas exchange organs.

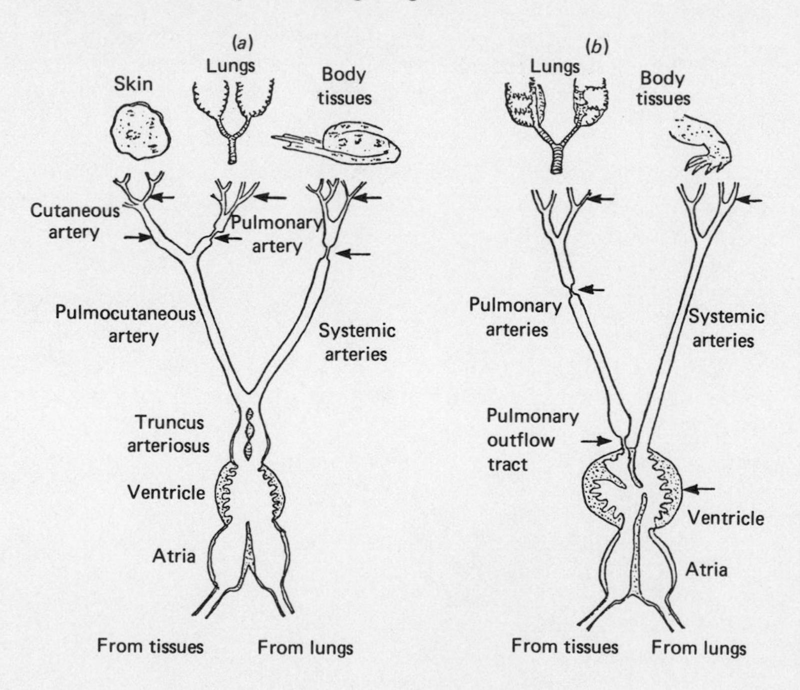

Tetrapods. The anatomical and physiological basis for the separation of bloodstreams in the undivided amphibian ventricle is not yet fully understood, but depends to varying degrees on the highly trabeculate inner ventricular surface, the complex arrangement of the specialized spiral valves in the conus arteriosus, and a laminar flow of blood in the heart (Shelton 1975). However, in all of the amphibians, oxygenated or partially oxygenated blood draining the skin is returned to the heart in the systemic venous circulation; so there is also an extracardiac admixture of blood tending to raise systemic venous P_{O_2}.

Whether the degree of separation of oxygenated and deoxygenated blood in the morphologically more divided amphibian circulation exceeds that of the Dipnoi, for example, is equivocal, and in any event comparison of this single variable will not necessarily reveal any "superiority" in terms of the total metabolic cost of exchanging respiratory gases between the tissues and the environment. There may be some hemodynamic advantage in the arrangement of the amphibian circulation, however, in that there is no intervening branchial vascular bed between the cardiac pump and the lungs. Thus the major exchange organs are perfused at a pressure very similar to ventricular pressure, and the total impedance from the heart across the gas exchange organ to the veins may be less than in most of the air-breathing fish. The exceptions among the amphibians are anuran tadpoles and some urodeles, which possess well-developed gills and lungs perfused in series.

As in other intermittent breathers, the onset of spontaneous lung ventilation in anuran amphibians is accompanied usually by a variable increase in cardiac output, mainly the result of an increase in stroke volume of the heart. Pulmocutaneous blood flow in the clawed toad *Xenopus laevis* may increase two to four times above prebreath levels within a few heartbeats of the start of ventilation, although there is often no accompanying change in systemic blood flow (Figure 3.13). Thus almost all of the increase in cardiac output is directed into the pulmocutaneous circuit. Such changes in the relative rates of perfusion of these two parallel circuits result primarily from striking fluctuations in the impedance of the pulmocutaneous circulation, with systemic vascular impedance remaining almost unchanged (Shelton 1970a). A large fall in impedance of the pulmocutaneous circuit occurs following breathing in *Xenopus*, as indicated by the decrease in mean pressure and increase in pulse pressure (Figure 3.13). The net result of these hemodynamic events in anurans is a preferential redistribution of cardiac output to the lungs and skin. Strong vasomotor activity has been demonstrated in vitro in both the cutaneous and pulmonary arteries of anuran amphibians (McLean and Burnstock 1967; Campbell 1971; Smith 1976). Changes in vasomotor activity in the systemic and lung circulations could alter not only the impedance balance between the systemic and pulmocutaneous circulations but also that between the lungs and the skin (Figure 3.12).

Unfortunately, few studies have attempted to determine directly how or if blood admixture in the ventricle is influenced by intermittent breathing. During periods of apnea, pulmonary and possibly cutaneous blood flow may fall to low levels (Figure 3.13), and a large right-to-left shunt must prevail. When pulmocutaneous flow rises and pulmonary gas exchange increases during and shortly after breathing, effective separation of systemic and pulmonary venous blood would be more desirable. In the urodele *Amphiuma*, Toews et al. (1971) have shown

Figure 3.13. Pressures and flows in the arterial arches of *Xenopus* (85 g). Pressure changes in the buccal cavity produced by movements of the buccal floor are recorded on the upper trace. Each of the bursts of movement recorded was of the lung ventilating type. The effect of breathing movements on individual flow and pressure pulses can just be seen (from Shelton 1970a).

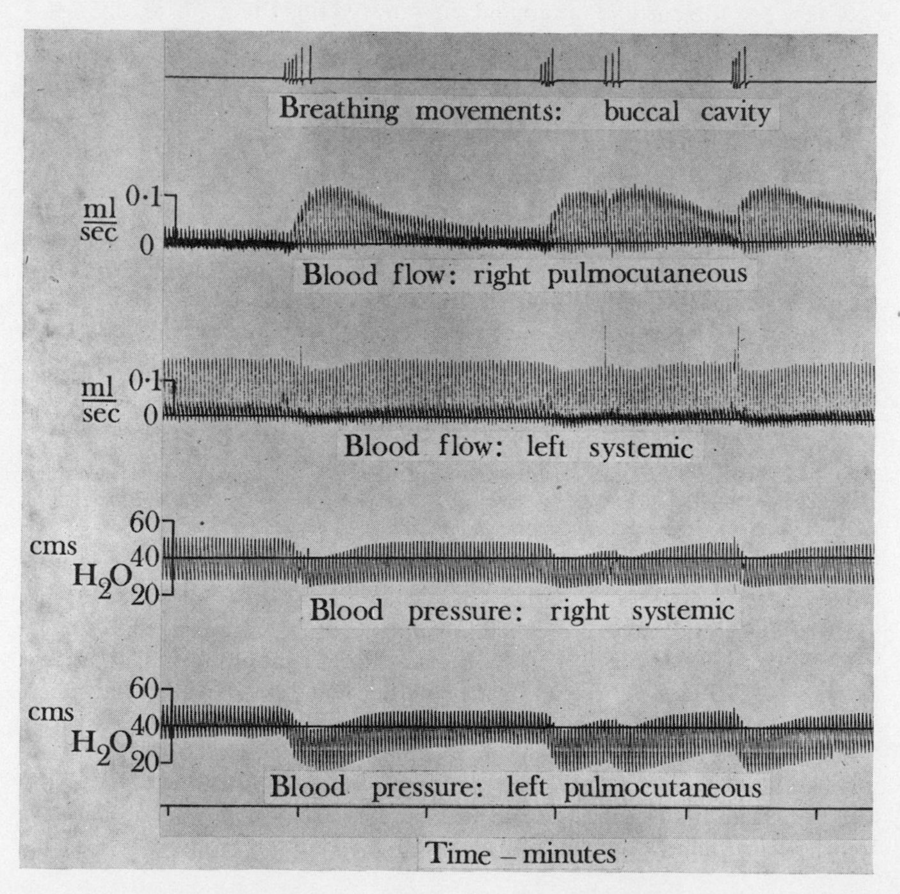

that during lung ventilation, when pulmonary P_{O_2} is highest, an average of 80% of blood perfusing the dorsal aorta is derived from the pulmonary veins. During very low lung P_{O_2}'s, however, less than 40% of dorsal aorta blood comes from the lungs. Thus, in this amphibian at least, there are not only changes in the absolute magnitude of blood flow correlated with intermittent breathing, but the composition of blood perfusing the arterial arches closely reflects the need for ventricular separation of arterial bloodstreams as influenced by active lung ventilation or periods of apnea.

The reptiles represent those vertebrates among the bimodal, intermittent breathers that possess the greatest anatomical division of the heart. The lungs, which are perfused by a discrete pulmonary circulation, are clearly the major respiratory organs because the skin plays a variable but usually very small role in the exchange of O_2 and CO_2. In addition to the large fluctuations in the heart rate of reptiles during intermittent breathing (Figure 3.3), there are profound readjustments in cardiac output and its redistribution to the lungs and the trunk that are clearly correlated with intermittent periods of lung ventilation (Figure 3.14). This is true not only for the Chelonia, but for reptiles in general (White 1976). A combination of increased heart rate and cardiac stroke volume with the onset of air breathing can raise cardiac output by 30–120% in *Chrysemys*. Flow to the pulmonary circuit, in particular, rises greatly during air breathing in aquatic or semiaquatic chelonians (White and Ross 1966; Shelton and Burggren 1976; Burggren et al. 1977), with an increase over prebreath values of 3 to 10 times in pulmonary artery minute flow common in *Chrysemys*. Systemic minute flow in the right aorta is usually affected to a much lesser extent, increasing on average by perhaps 50% during lung ventilation. Thus, 60–65% of the cardiac output perfuses the lungs (a net left-to-right shunt) during breathing, but after periods of apnea longer than 5 min in *Chrysemys*, pulmonary perfusion may fall to only 45% of cardiac output (a net right-to-left shunt).

Lung ventilation is more frequent, and hence interbreath intervals are much shorter, in many terrestrial reptiles compared to many semiaquatic forms (see Figure 3.3). Consequently, the ventilatory convective flow oscillates to a lesser extent, and pulmonary perfusion and cardiac output, as well as lung and blood gas levels, are more stable in the terrestrial tortoise *Testudo* compared to the aquatic turtle *Chrysemys* (Shelton and Burggren 1976; Burggren et al. 1977).

The Crocodilia and the varanid lizard (Webb et al. 1971; Millard and Johansen 1974) are the only reptiles with a separation of the ventricle into a dual pressure pump. The heart and ventricular outflow tracts of the remaining noncrocodilian reptiles are considerably more complex in arrangement than in the amphibians; however, the lack of mor-

Figure 3.14. Effect of lung ventilation and accompanying tachycardia on sytemic and pulmonary blood pressure and flow in an unrestrained *Chrysemys*. At the start of the records the turtle was voluntarily diving. The turtle surfaced (first arrow), then began a short period of lung ventilation (second arrow). This was followed after 4 min by another short period of lung ventilation (third arrow). The flow records were used to derive minute flow as plotted in the graph. Closed circles indicate left pulmonary minute flow, open circles right aorta minute flow. (At very low heart rates the pulmonary flow probe shows some zero drift due to reduced vessel size.) (From Shelton and Burggren 1976.)

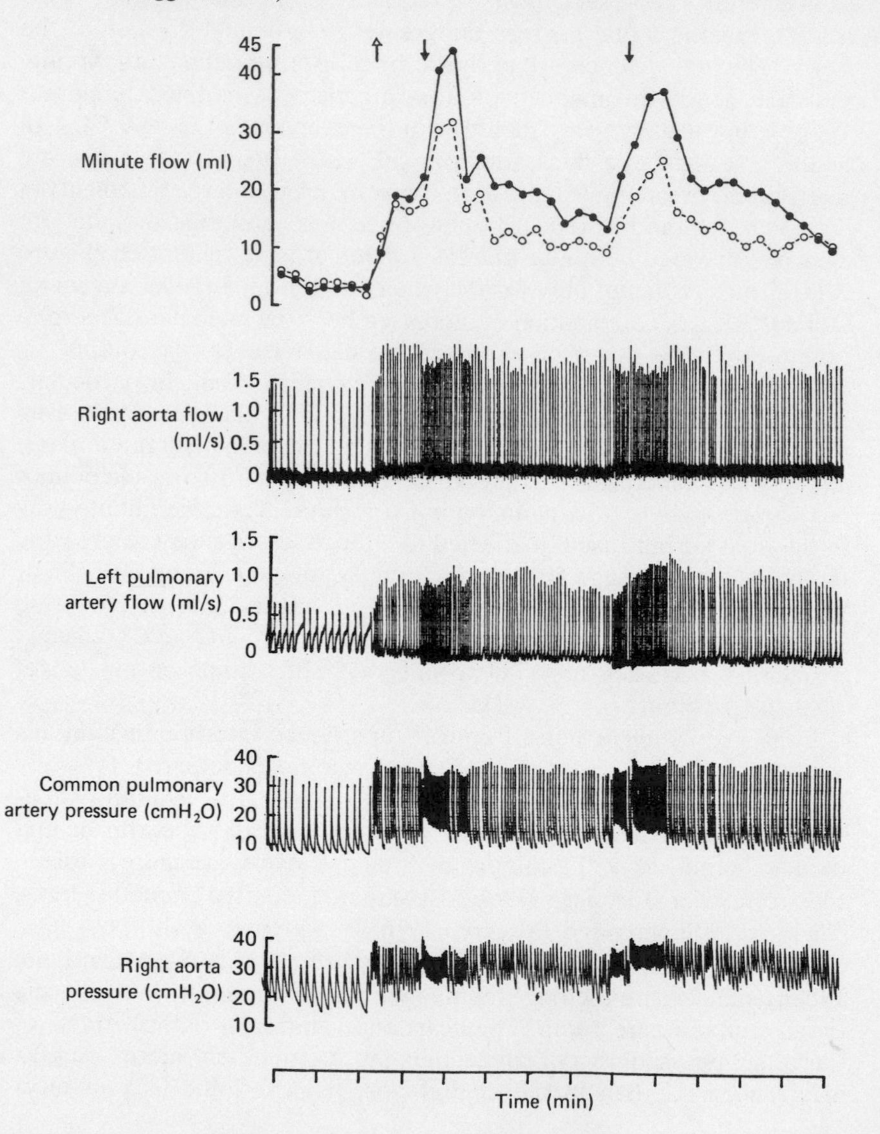

phological and functional division of the circulation to the lungs and that to the trunk dictates that some intracardiac shunting must always occur (Shelton and Burggren 1976). Early investigations (see White 1976), utilizing dye techniques or blood sampling, demonstrated the general noncrocodilian condition to be one of substantial, though not complete, separation of deoxygenated and oxygenated blood within the structurally complex ventricle. Direct measurements of pulmonary and systemic blood flow and blood gases in unanesthetized, unrestrained turtles (White and Ross 1966; Shelton and Burggren 1976) have shown that both left-to-right and right-to-left intraventricular blood shunting occurs, and thus the composition of blood conveyed in the pulmonary and systemic arteries is variable. In *Chrysemys scripta,* only about 40% of the blood entering the right aorta during periods of apnea is oxygenated blood from the left auricle, but after the onset of breathing nearly 70% of blood conveyed by the right aorta is oxygenated. Quite consistently, 90% of deoxygenated blood from the right auricle enters the pulmonary artery throughout intermittent breathing, in spite of the large fluctuations in pulmonary blood flow and the net intracardiac shunt (Burggren and Shelton 1979), thus facilitating O_2 and CO_2 transfer in the lungs between breaths.

In the noncrocodilian reptiles, then, lung ventilation is accompanied by an increase in pulmonary perfusion, an improved separation of venous blood in the heart, and consequently an augmented O_2 delivery to the tissues and CO_2 delivery to the lungs. During apnea, although the lungs still receive largely deoxygenated blood from the tissues, total pulmonary perfusion is reduced, and only a small amount of oxygenated blood is circulated to the systemic tissues. If the periods of apnea are long, and lung gas P_{O_2} decreases considerably, then the P_{O_2} difference between right and left atrial blood will be progressively reduced. Separation of the large systemic venous inflow (65% or more of cardiac output) from the small pulmonary venous inflow thus becomes less critical during prolonged apnea, and neither *Testudo* nor *Chrysemys* sustains a selective perfusion to the systemic circulation under these conditions.

In most reptiles, as in amphibians, a change in the impedance balance between body and lungs is the major way in which a redistribution of cardiac output can be achieved. During lung ventilation in chelonian reptiles, for example, systemic impedance remains relatively constant, but the impedance of the pulmonary circuit falls by one-third (Shelton and Burggren 1976), establishing a pattern of impedance variation similar to that in many amphibians (Figure 3.12). In addition to putative vasomotor responses in the reptilian lung circulation (White and Ross 1966), strong vasomotor activity can develop in the extrinsic pulmonary arteries (Berger 1972; Burggren 1977a; Smith and Macintyre 1979) or even in the pulmonary outflow tract of the ventricle (Burggren 1977a,b; Smith and Macintyre 1979). Additionally, intracardiac blood shunting, at least in chelonian reptiles, can be influenced to some extent by active, ventilation-correlated changes in the beat-to-beat activation pattern of the cardiac muscle, producing fluctuations in the rate and timing of contraction of the various cardiac chambers (Burggren 1978). Hence, the amount of pulmonary perfusion can be influenced by vasoconstriction or vasodilation at any one of three regions in the pulmonary circuit, and an additional intracardiac mechanism can influence the composition of the blood conveyed to the lungs and the trunk.

To summarize, continued perfusion of the aerial gas exchanger during prolonged periods of apnea can be metabolically expensive relative to any gas exchange it achieves. Thus, almost all of a wide variety of bimodal and/or intermittent breathers carefully adjust perfusion of their suprabranchial chambers, gas bladders, or lungs to maximize O_2 extraction and CO_2 elimination and to minimize oscillations in O_2 delivery to the tissues when these organs are not being ventilated. The mechanisms behind these adjustments in perfusion of the gas exchanger are complex and varied, but usually involve changes in cardiac output as well as blood flow redistribution between the circulation of the gas exchanger and the body tissues.

The transition from water to air breathing has seen fundamental rearrangements in the cardiovascular system. In the ancestral aquatic breathers all cardiac output perfused the gas exchange organ, but the progressive exploitation of bimodal breathing necessitated an increasing elaboration and separation of the circulation. The tetrapods ultimately abandoned any need for aquatic exchange of gases or ions by transferring these processes to lungs and kidneys, respectively (Chapter 2). This change permitted a return to the original conditions with all cardiac output perfusing a single gas exchange organ, with the transitional circulatory arrangements condensed into a divided circulation laid out in series. Concomitant with the evolution of these cardiovascular modifications necessary for bimodal breathing were fundamental rearrangements in the mechanisms by which air was supplied to the respiratory membranes. These mechanisms will now be examined.

4 *Mechanisms of ventilation*

Very different modes of ventilation exist for lungs and gills. In fish a pressure gradient drives a unidirectional flow of water across the gills from the buccal cavity to the opercular cavities (Figure 4.1a). The synchronized action of a buccal pump and an opercular pump generates the pressure gradient. Mammalian lungs, however, are tidally ventilated by aspiration. Below-ambient pressures are developed within the lung through expansion of the thoracic cage to draw air into the lung (Figure 4.1b). Thus the evolution of air breathing in vertebrates has been associated with a change from a unidirectional mode of ventilation of a respiratory surface with water to one with a tidal mode of ventilation with air using aspiration. Clearly these changes are dictated by the marked differences in density and viscosity between air and water (see Chapter 2).

Most bimodally breathing vertebrates use a modified buccal force pump to ventilate their air-breathing organ. The modified buccal force pump is analogous to, if not homologous to, the buccal pump used to ventilate the gills. The positive-pressure phase of the buccal pump is utilized to force-ventilate the air-breathing organ. These three major forms of vertebrate ventilation are shown schematically in Figure 4.2 and Table 4.1.

To examine the evolution of air-breathing mechanisms, we must first consider gill ventilation in fish breathing water, particularly the nature of the buccal force pump. We then can see how buccal force pumping was modified for air breathing, and why it was subsequently replaced by aspiration breathing.

Gill ventilation

Water is forced over the gills of bony fish by alternate expansion and contraction of the buccal and opercular chambers. This is achieved by the actions of muscles in the buccal and opercular wall. Each cycle of the buccal and opercular pumps consists of an expansion and a compression phase. In trout the buccal cavity is expanded laterally by muscle action, the buccal floor drops as the mouth opens (Figure 4.3), and water is drawn into the buccal cavity. The mouth then closes and/or is sealed by the buccal flap (a one-way valve), and lateral compression of the buccal cavity creates a pressure above that in the opercular chamber. This represents the cycle of the buccal force pump and is repeated between 10 and 90 times per minute, depending on the species and physiological state of the animal. While the buccal pump expands laterally, the opercular cavities are also enlarged by opercular adduction. A phase of subambient opercular pressure develops because the opercular flaps remain in contact with the cliethrum. Because opercular pressure is less than buccal pressure (Figure 4.3), water will flow

Figure 4.1. (*a*) An internal view of the buccal and opercular cavities in the head of a teleost fish. The unidirectional flow of water from the buccal cavity to the opercular cavities on either side of the head and over the eight gill arches is indicated. (*b*) A schematic representation of a mammalian thoracic cavity showing the respiratory structures. The tidal air flow into each lung is indicated.

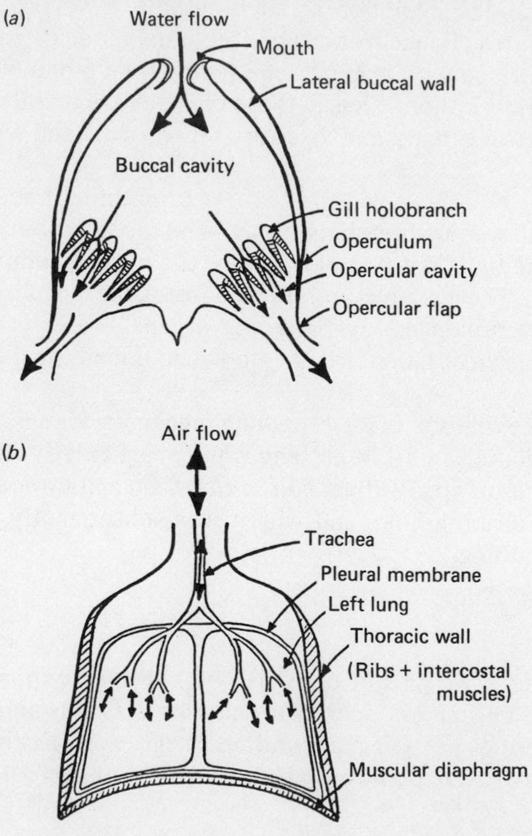

across the gills, in spite of below-ambient pressures on either side of the gills. Opercular adduction is synchronized with buccal compression, and water is exhaled through the open opercula. An essentially continuous flow of water is maintained across the gills, even though water flow is discontinuous into the mouth and out of the opercula. Water flow across the gills does oscillate, however, and the greatest water flow occurs during buccal compression (Holton and Jones 1975).

Synchrony between the buccal and opercular pumps is essential to maintain the positive differential pressure gradient across the gills. Phasic contraction of muscle groups of both respiratory pumps is clearly evident (Figure 4.4). However, two other factors are extremely important in achieving synchrony:

1. Single muscle groups act on both pumps. The buccal force pump involves lateral movement of the flexible palatal complex. Adduction of the palatal complex (buccal compression) is produced by contraction of the adductor arcus plantini et operculi. As the name suggests, this muscle also adducts the opercula, to allow exhalation.

2. Skeletal coupling is also important. The units of the teleost skull are intimately connected by ligaments, such that movement of one skeletal unit is transmitted to others. For example, raising of the lower jaw to close the mouth (adductor mandibulae contraction) also raises the hyoid, laterally adducting the hyomandibula and, with it, the operculum. Similarly, raising of the buccal floor through hyohyoideus contraction against the hyoid also results in a folding of the gill arches.

Table 4.1. *The distribution of the three principal ventilation mechanisms among vertebrates, including usual environment.*

Ventilatory mechanism	Respiratory medium	Vertebrate group	Habitats
Buccal and opercular pumps with unidirectional flow of medium	Water Air	Fish (gills) Some air-breathing fish (anabantids)	Aquatic Aquatic
Modified buccal force pump with tidal flow of medium	Air	Most air-breathing fish Dipnoi Amphibians	Aquatic
Aspiration pump with tidal flow of medium	Air	*Arapaima* Reptiles Birds Mammals	Aquatic Majority terrestrial but some aquatic forms

Respiratory water flow is increased by increases in both ventilation rate and stroke volume. To achieve a greater stroke volume more muscle groups are involved in ventilation (Ballintijn and Hughes 1965). In many fish, increases in water flow are achieved by greater emphasis on buccal pumping, the cavity being greatly enlarged by ventral buccal floor depression. Active compression of the buccal floor during hyperventilation produces a formidable force pump, and much higher buccal pressures are generated. Buccal pressure oscillations exceeding 15 cmH_2O of water are not uncommon with deep ventilation in ling cod, whereas at rest the pressure oscillations hardly reach 1 cmH_2O.

Figure 4.2. A schematic representation of the three principal mechanisms of ventilation in vertebrates. Arrows indicate direction of water or air flow. (*a*) Synchronous buccal pump (BP) and opercular pump (OP) moving water unidirectionally from the buccal cavity (BC) to the opercular cavity (OC). (M) mouth or nares and associated flap or sphincter; (O) operculum and associated opercular flap. (*b*) The modified buccal force pump (MBP) used to tidally ventilate air-breathing organs (ABO). The flow to the sac is sphincter (SPH)-controlled. (*c*) The aspiration pump (AP) used in tidal lung (L) ventilation. (D) diaphragm; (TC) thoracic cavity; (AC) abdominal cavity.

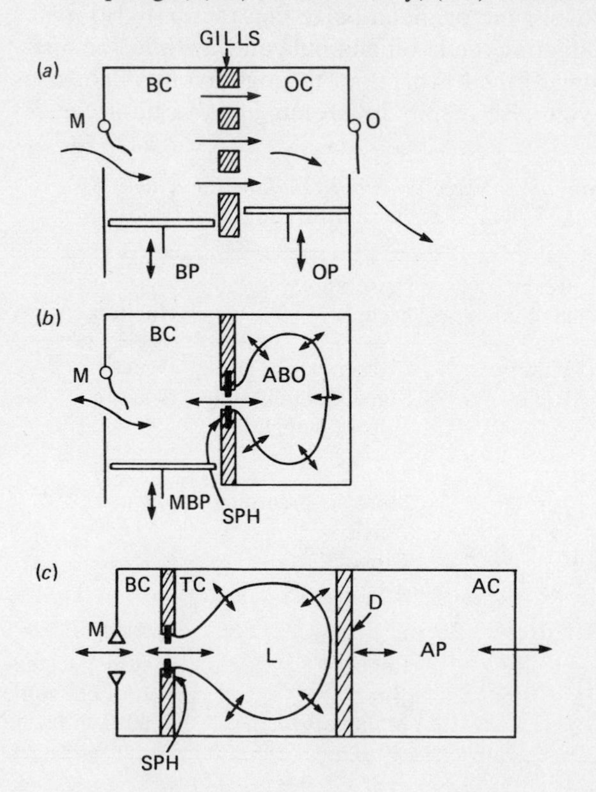

The buccal or opercular pumps are not always of equal importance in generating flow. The dominance of one pump or the other in a given species is related to its ecological situation (Baglioni 1907; Hughes 1960). Among the pelagic and actively swimming fish the buccal pump is usually dominant (e.g., horse mackerel), or both pumps are of equal size (e.g., trout, wrasse, and herring). The opercular pump takes on a greater significance in bottom-dwelling fish. In such cases, the opercular movements are large, whereas buccal movements are hardly perceptible at rest (e.g., ling cod). In other teleosts, the opercular cavity becomes more compliant, and the opercular openings are small (e.g., *Conger conger* and *Callionymus*). In many fish, gill ventilation is

Figure 4.3. Trout (70 g). The breathing movements of the mouth and operculum, together with associated pressure changes in the buccal and opercular cavities. The differential pressure between these cavities is shown in the bottom trace. O and C indicate the opened and closed positions of the mouth, operculum, and their associated valves (from Hughes and Shelton 1957).

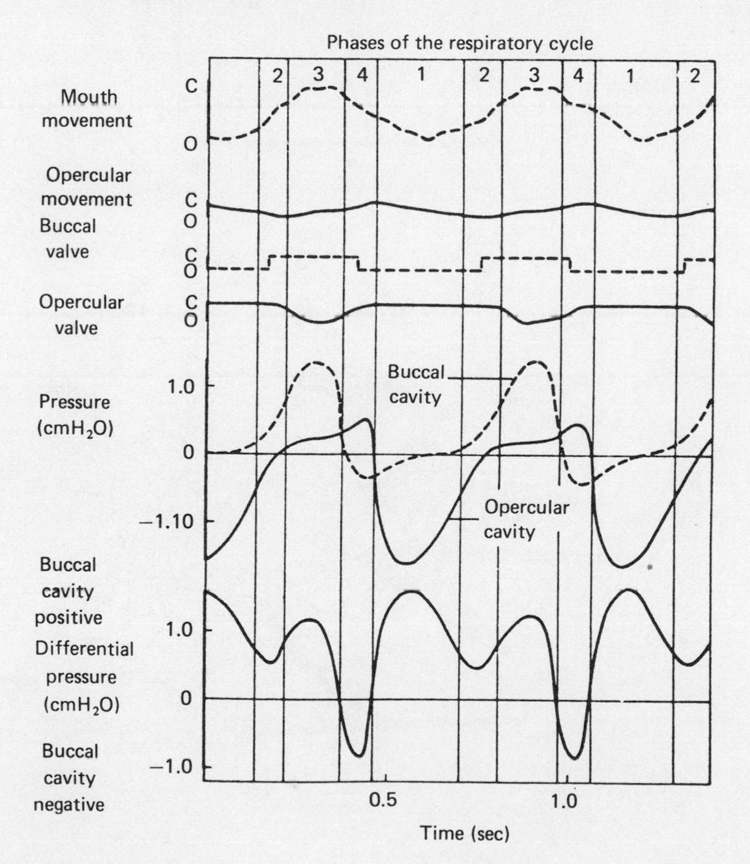

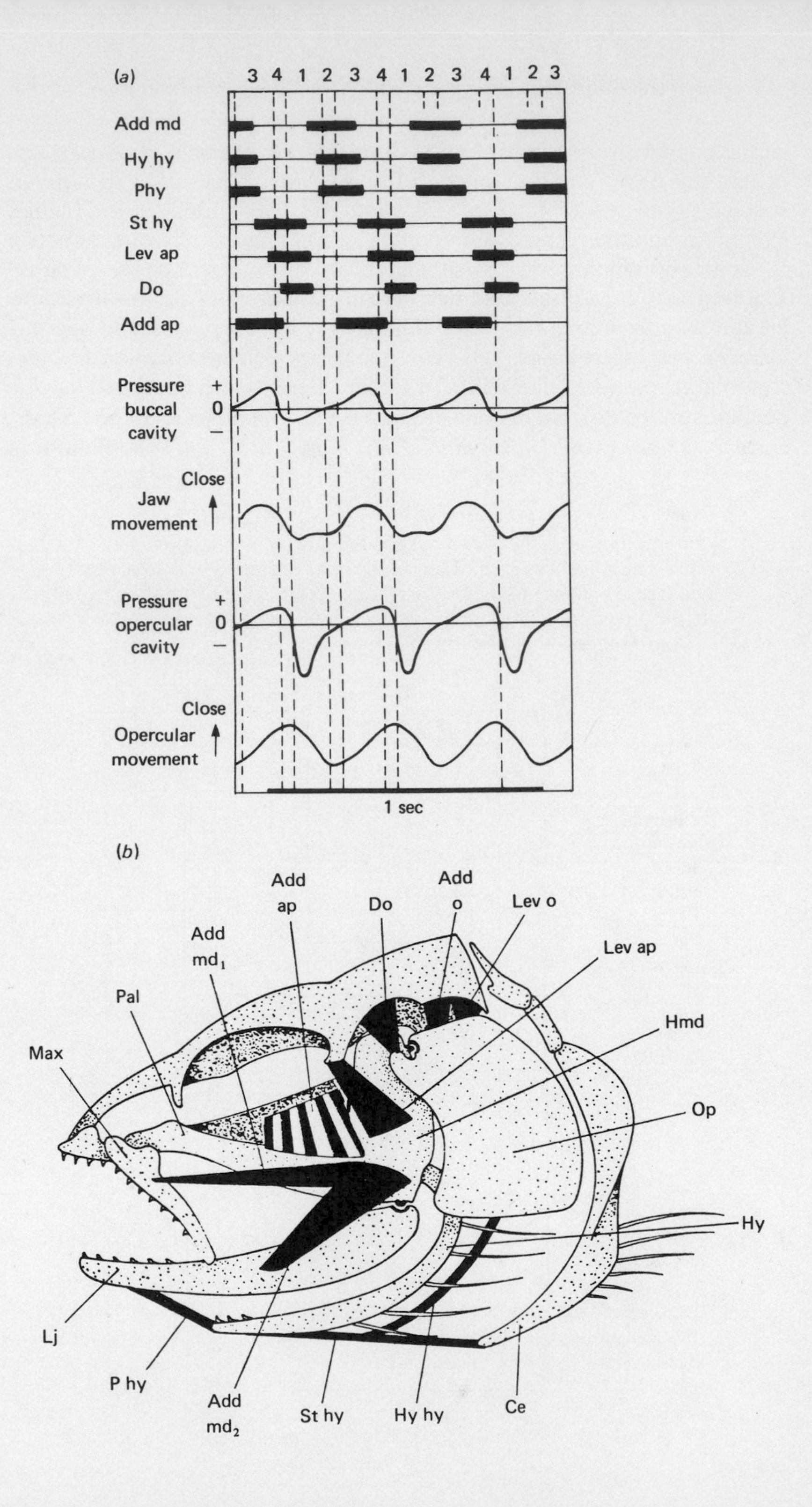
(a)
3 4 1 2 3 4 1 2 3 4 1 2 3
Add md
Hy hy
Phy
St hy
Lev ap
Do
Add ap
Pressure buccal cavity
+
0
−
Close
Jaw movement
Pressure opercular cavity
+
0
−
Close
Opercular movement
1 sec
(b)
Add ap
Do
Add o
Lev o
Add md1
Lev ap
Pal
Hmd
Max
Op
Hy
Lj
P hy
Add md2
St hy
Hy hy
Ce

achieved also by forward motion through the water with the mouth open.

Modifications of the buccal pump for air breathing

The majority of air-breathing fish are, in fact, teleosts. Gill ventilation, whenever present, is essentially the same as in strictly aquatic breathers. Often a modified buccal pump is used also to ventilate the air-breathing organ. The nature of this modified buccal force pump is similar for air-breathing fish that respire with a gas bladder, (e.g., gar, *Lepisosteus osseus;* Rahn et al. 1971), the stomach, (e.g., *Hoplosternum littorale;* Carter and Beadle 1931), or elongated pharyngeal air sacs, (e.g. *Saccobranchus fossilis;* Das 1927), and for Dipnoi and anurans that respire with lungs.

Inhalation of air is accomplished by buccal expansion and compression. The buccal movements are exaggerated in a ventral rather than lateral direction and achieve the maximal or near-maximal buccal volume. The buccal floor movements are analogous, if not homologous, to those used for ventilation of the gills with water. A comparison of the activity of respiratory muscles during air breathing and water breathing illustrates this point. Muscle activity and the pressures generated during air and water breathing are shown in Figure 4.5, and can be compared with muscle activity and pressures associated with gill ventilation in a water-breathing teleost (Figure 4.4). The first point to notice is that the activities of muscles associated with ventilation of the gills with water are similar in both lungfish and teleosts. The modifications associated with the use of the buccal pump for air breathing are surprisingly simple. First, in water breathing the buccal pump is used to develop a continuous flow of water over the gills, whereas in air breathing it is used simply to force air into the gas bladder or lung. Thus during air breathing the muscles of the buccal floor and lower jaw are all active together at the onset of buccal compression (Figure 4.5b,c). This situation can be compared with the delay between the contraction of muscle

Figure 4.4. (*a*) Breathing movements and pressure changes in the buccal and opercular cavities of the trout, together with a diagrammatic representation of the time relations of activity in seven respiratory muscles (after Ballintijn and Hughes 1965). (*b*) Lateral view of the skull and important respiratory muscles in a typical teleost. (Max) maxilla; (Pal) palatine; (Add md_1 and md_2) adductor mandibulae (two divisions); (Add ap) adductor arcus palatini; (Do) dilator operculi; (Add o) adductor operculi; (Lev o) levator operculi; (Lev ap) levator arcus palatini; (Hmd) hyomandibula; (Op) operculum; (Hy) hyoid; (Ce) cleithrum; (Hy hy) hyohyoideus; (St hy) sternohyoideus; (P hy) protractor hyoidei; and (Lj) lower jaw (from Shelton 1970b; reprinted by permission of Academic Press, Inc., which holds the copyright).

groups during gill irrigation (Figure 4.5a), where the buccal pump is being used to create a more even flow of water.

A second modification for air breathing is the uncoupling of movements of the buccal and opercular pumps. In water breathing, lateral buccal compression is associated with an opening of the opercular cleft. This is clearly a liability during air breathing, as air will escape via the open opercular cleft rather than be forced into the air-breathing organ. The selection of ventral rather than lateral buccal movement during air breathing reduces the interaction between buccal and opercular pumps. The palatal complex, which is involved in coupling lateral buccal movements to opercular opening in water-breathing teleosts, is fused to the skull in lungfish. To prevent air loss via the opercular cleft during buccal compression, the slit is closed during air breathing by contractions of the hyoideus muscle in *Protopterus* and probably in many other air-breathing fish, especially during buccal compression (Figure 4.5b). In many air-breathing fish the opercular slits are also reduced in size or fused to form a single ventral opening (Carter and Beadle 1931). In some species the operculum is sealed to the cleithrum by mucus during air breathing; for example, in *Tomidon* (Johansen 1970) and *Pseudoapcrytes lanceolatus* (Das 1934).

On land the opercular pump is clearly redundant, so it is not surprising that it is absent in frogs. In addition the lateral skull bones are reduced in forms such as *Saccobranchus,* lungfish, and anurans. The selection of light jaw structures is correlated with the increased work load in moving heavy skeletal structures during ventilation. Thus anurans lack an opercular cleft, have a much reduced buccal and jaw bone structure, and possess a fused palatal complex.

The synchronous contraction of buccal and jaw muscles during air breathing results in the generation of high buccal pressures. This is perhaps surprising because the viscosity of air is low compared with water, and one might expect, therefore, lower buccal pressures during air breathing. Clearly the high buccal pressures must reflect a resistance to lung inflation, which in turn may result from two factors, namely, a high resistance to flow in the air ducts and a high viscoelastic resistance to increases in volume of the gas bladder or lung. Swim bladders, for instance, must retain a constant volume under a variety of pressures to function in buoyancy control. In general, swim bladders are not very compliant and are limited by a stiff body wall. Air bladders and lungs are far more compliant; they deform readily and enlarge with increased pressure. The body walls, however, restrict air bladder or lung inflation, particularly at higher lung volumes (Figure 4.6). The relative contributions of the body wall and air-breathing organ to the total viscoelastic resistance are difficult to determine. The viscoelastic resistance of the body walls is, however, apparently less in anurans

Figure 4.5. A comparison of mechanisms for lungfish gill and lung ventilation and anuran lung ventilation. (*a*) The aquatic cycle of gill ventilation in *Protopterus*. Continued.

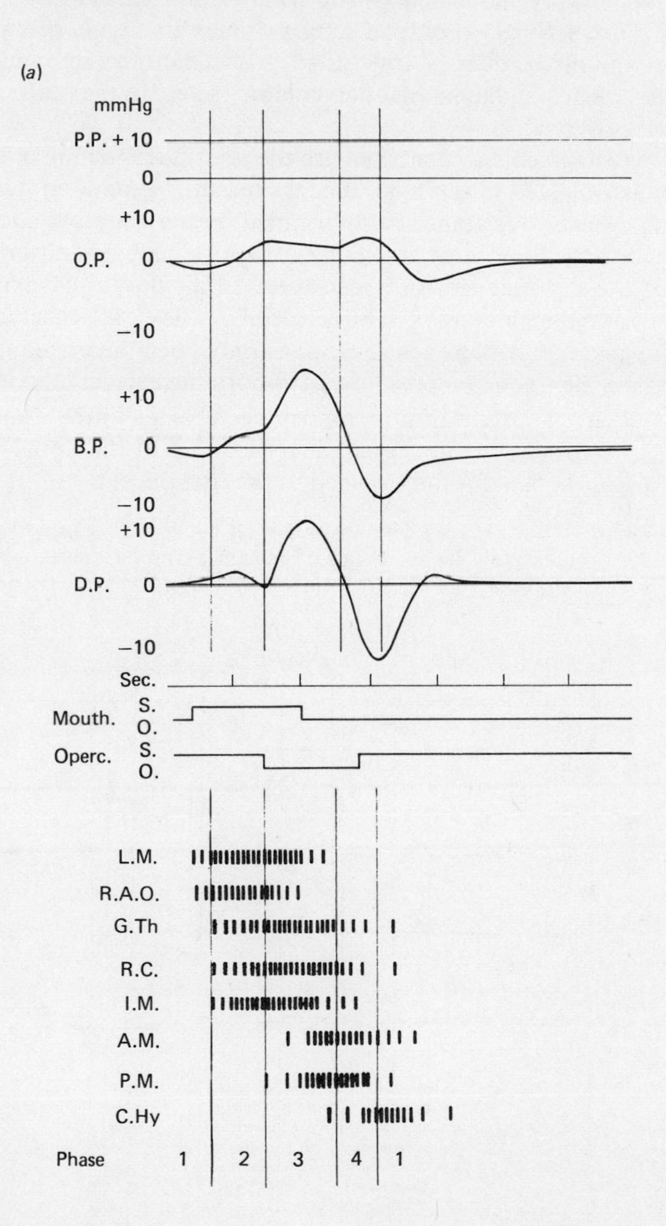

compared to fish and lungfish. Anurans have reduced ribs and thoracic muscles that increase body wall compliance such that the wall easily follows ventilatory movements. The heavy flank muscles of fish and lungfish must set rigid limits on lung volume. In *Saccobranchus,* for instance, the air bladder is embedded in the flank musculature (Das 1927), and clearly maximal bladder volume is set by the nature of the muscular body walls.

Airway resistance may contribute to the resistance to lung or bladder inflation as reflected in the high buccal pressures generated during air breathing. Airway resistance is determined by the diameter and length of the duct and the rate of air flow. Reduced diameter and increased length of the air duct increase resistance to air flow, but changes in diameter have the more marked effect. High air flow rates also increase airway resistance. Airway resistance has only been measured in mammals, where the trachea represents an important resistance to air flow. There is clearly a large variation among vertebrates in the dimensions and velocity of air flow in the air ducts leading to the aerial gas exchange organ. In fish this duct is used to prevent the entry of water into

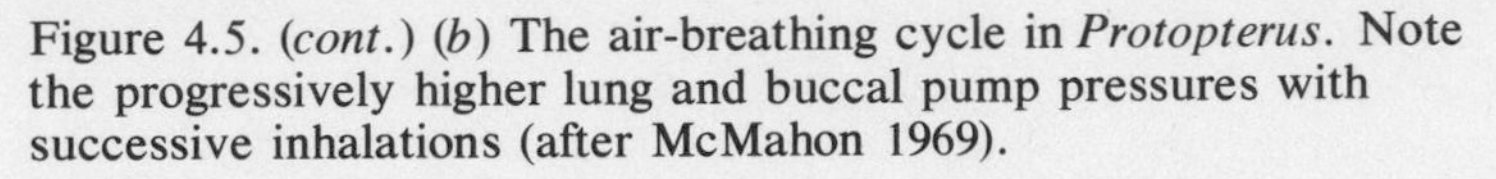

Figure 4.5. (*cont.*) (*b*) The air-breathing cycle in *Protopterus*. Note the progressively higher lung and buccal pump pressures with successive inhalations (after McMahon 1969).

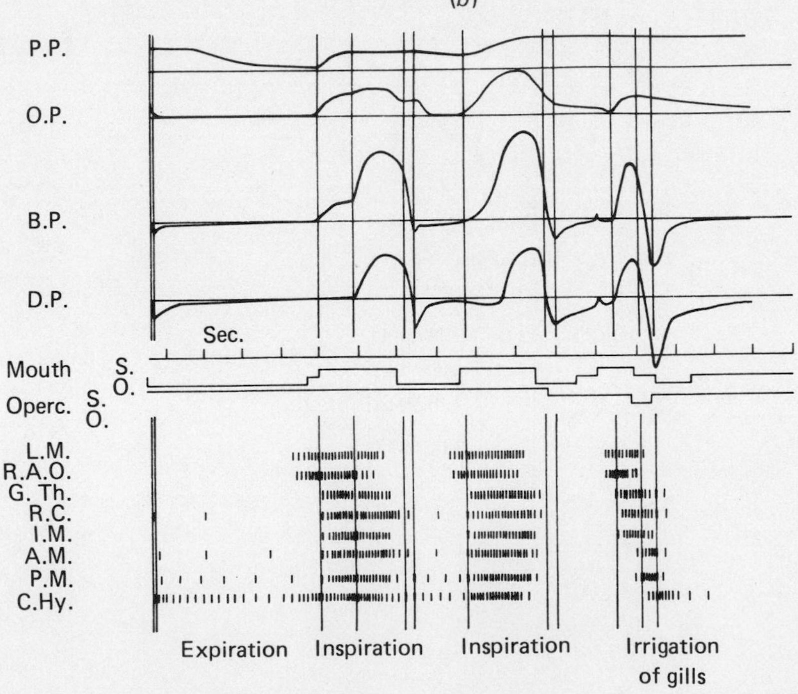

Figure 4.5. (*c*) Semischematic diagram of pressure and volume changes in a buccal and a lung ventilation cycle of a frog. In the volume trace, up is an increase in lung volume but a decrease in buccal volume; solid bars, muscular activity in all frogs investigated; broken bars, muscular activity in some frogs in this phase (from West and Jones 1975a). Abbreviations: (P.P.) intrapulmonary pressure; (O.P.) opercular pressure; (B.P.) buccal pressure; (D.P.) differential pressure between buccal and opercular cavities; (L.V.) lung volume; (B.V.) buccal volume; (O) open; (S) shut. Activity in: (L.M.) levator mandibulae muscle; (R.A.O.) retractor anguli oris muscle; (G.Th.) geniothoracicus muscle; (R.C.) rectus cervicis; (I.M.) intermandibularis muscle; (A.M.) anterior muscle of cranial rib; (P.M.) posterior muscle of carnial rib; (C.Hy.) constrictor hyoideus muscle. (Reproduced by permission of the National Research Council of Canada from the *Canadian Journal of Zoology* 53: 332–44, 1975.)

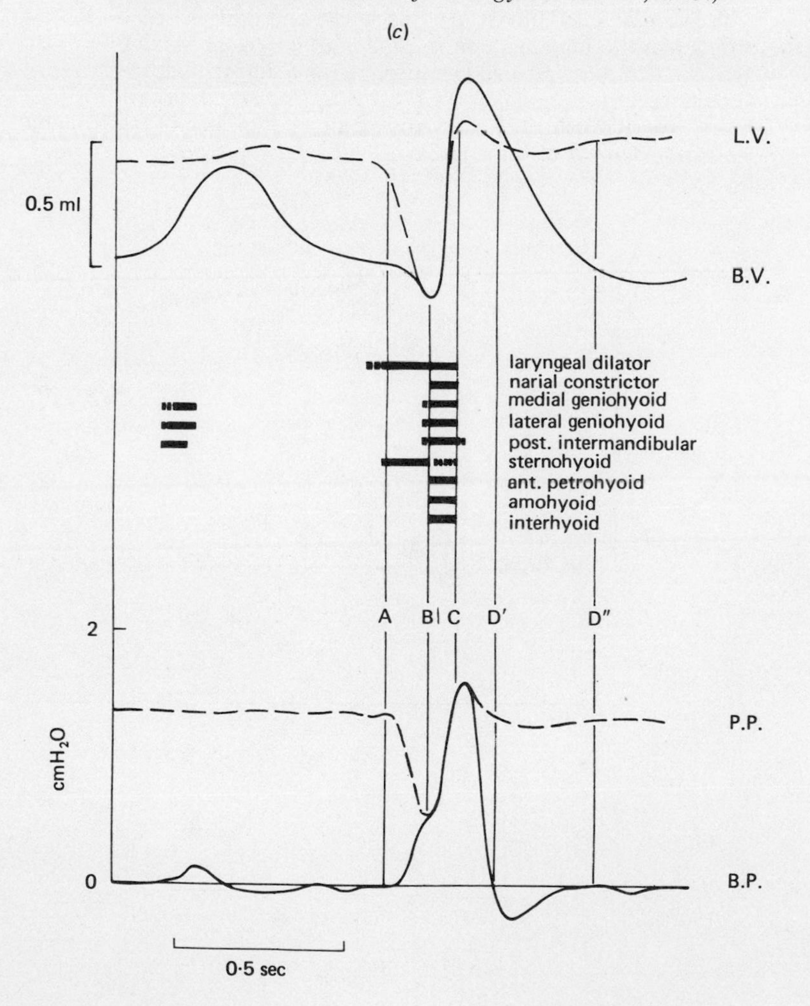

the bladder and the escape of air while the fish is submerged, as well as acting as a duct for air flow during ventilation. Often these fish can spend only a brief period at the water surface in order to reduce the chances of predation. Under these circumstances breathing must be rapid, and air flow may be high through a narrow air duct. Thus in many air-breathing fish, resistance to air flow in the pneumatic duct may be an important factor in the resistance to gas bladder inflation and may be a factor in the high pressures generated during buccal compression.

Although evidence is lacking, it seems probable that, in many fish and amphibians, exhalation is passive, especially when the animal is

Figure 4.6. The progressive elevation in lung and air bladder pressures with successive inflations. (*a*) Buccal (B) and lung (L) presures during a single lung inflation in *Bufo marinus* (from Macintyre and Toews (1976). (*b*) Intra-air-bladder pressure during multiple ventilation in two separate air breaths in *Hoplerythrinus* (from Farrell and Randall 1978). (Reproduced by permission of the National Research Council of Canada from the *Canadian Journal of Zoology* 54: 1364–74, 1976, and 56: 939–45, 1978.)

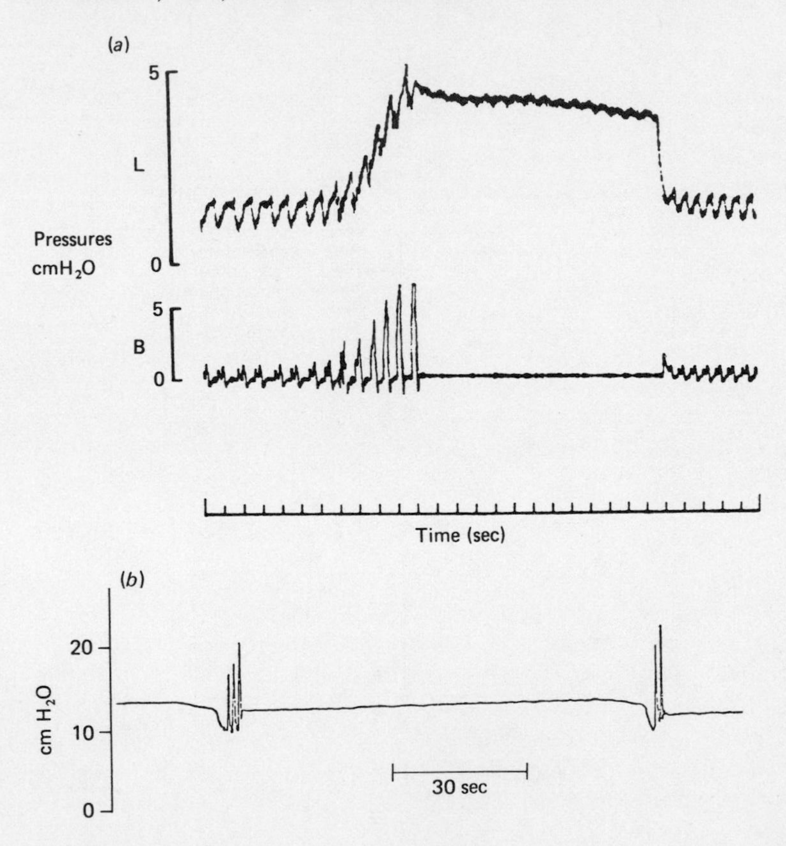

submerged, and intrapulmonary pressure is increased by hydrostatic pressure. In some instances hydrostatic pressure is sufficient to cause exhalation because the animal is at an angle to the surface (Figure 4.7). In air-breathing fish and Dipnoi hydrostatic pressure results in an intrapulmonary pressure that is about 10 cmH_2O above air pressures. In surfacing frogs the pressure difference between submerged lungs and the aerial environment is about 5 cmH_2O (Figure 4.7). In the jeju, *Hoplerythrinus,* inhalation precedes exhalation, which occurs as the animal swims down the water column. Only inhalation occurs at the surface, and it results in a high bladder pressure that, along with hydrostatic pressure, serves to cause rapid exhalation as the fish swims away from the surface (Figure 4.8). Active expiration may contribute to exhalation in lungfish and amphibians, particularly during the initial phases of expiration (Figure 4.9). Some air-breathing fish have smooth and/or striated muscles covering the outer surface of the air-breathing organ (e.g., *Saccobranchus, Amia,* and *Amphipnous*). These muscles undoubtedly aid exhalation, at least in *Amia*, where exhalation is rapid and violent and blows water out of the buccal cavity as the animal surfaces.

There are problems associated with using a modified buccal pump on land because hydrostatic support is no longer available. As a result, heavily structured jaws are uncommon in terrestrial air-breathing vertebrates using a buccal pump. Tetrapods use their legs and air-breathing fish their pectoral fins, to raise their ventrum off the ground to allow free movement of the buccal floor during buccal pumping. In fish, air bladder pressure and the stiff body walls probably prevent air bladder collapse under the fish's own weight in air.

The lungfish, *Lepidosiren* and *Protopterus*, and the bowfin, *Amia*, aestivate during periods of drought. Lungfish aestivate in a protective cocoon in the mud on river banks. The ventilatory pattern is changed as the animal forms the cocoon, tidal volume is reduced because of the restrictive nature of the cocoon, and lung ventilation falls as the requirements for oxygen are markedly reduced during aestivation (Figure 4.9).

Unidirectional air flow using a buccal pump

A few air-breathing fish, particularly the Anabantidae, ventilate their air-breathing organ with a unidirectional air flow (Figure 4.10). In the anabantoids both the opercular and buccal pumps are involved in air ventilation. The anabantoid air-breathing organs are located dorsal to the opercular cavity in paired, rigid, suprabranchial chambers. Because the inhalant openings are located in the pharyngeal cavity and the exhalant openings are in the opercular cavity, the direction of air ventilation parallels gill ventilation in most members of this family. Thus, the

Figure 4.7. The importance of a hydrostatic pressure during air breathing to drive passive exhalation in the aquatic environment. (*a*) The gourami. (*b*) *Callamoicthys*.

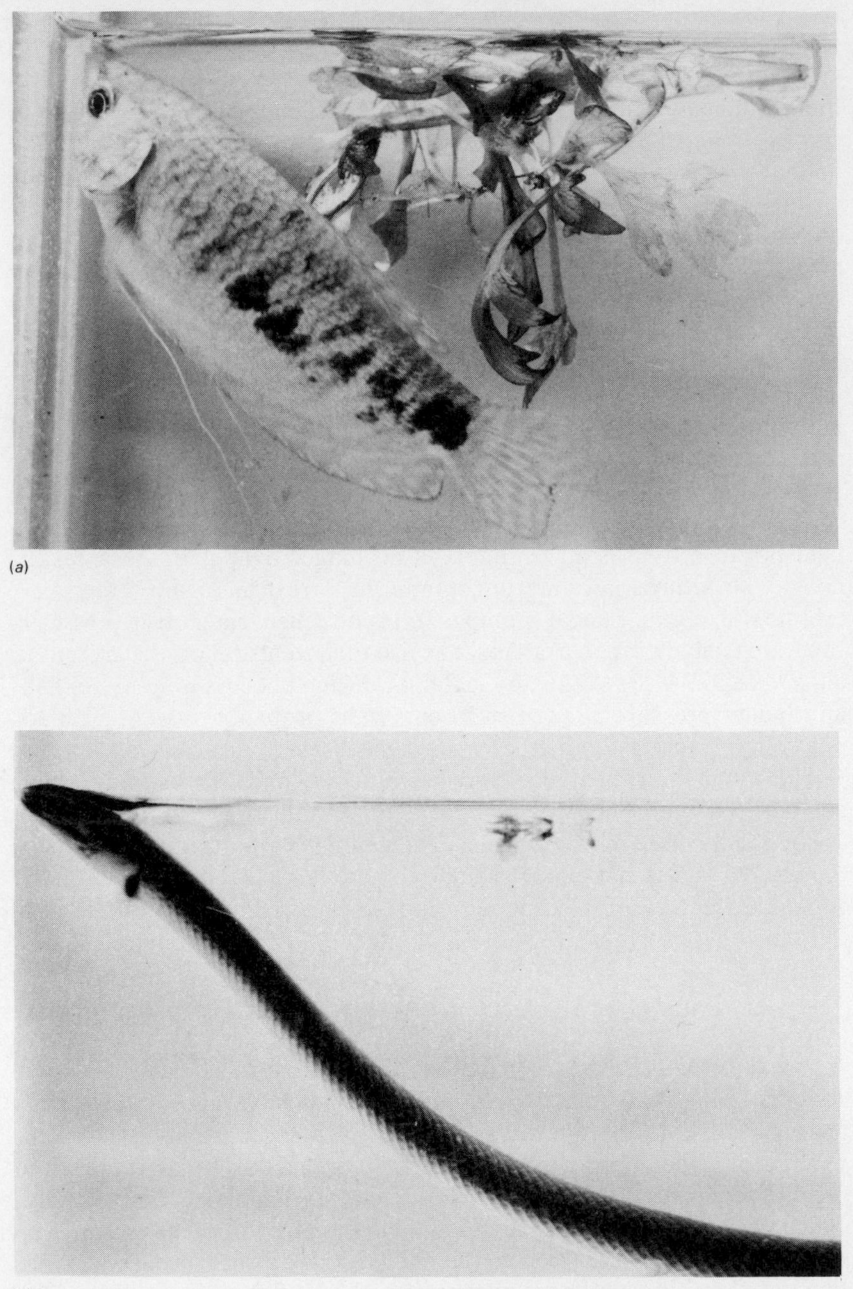

(*a*)

(*b*)

Figure 4.7. (*c*) The frog. The vertical line represents the approximate hydrostatic gradient from the location of the air breathing organ. Figure (*c*) from *Larousse Encyclopedia of Animal Life,* published by Paul Hamlyn Limited, London, New York, Sydney, Toronto (1976); photographed by T. Davidson.

(*c*)

unidirectional ventilation of the air sac, as found in the adult climbing perch *Anabas testudineus,* involves the same mechanisms as gill ventilation, but with air instead of water. Air is retained in the sac between air ventilations and during gill ventilation. The mechanical changes associated with air breathing are related to excluding water from the air chamber and directing air through the air chamber rather than through the gills. Neither problem has been examined experimentally. The exhalant openings are protected on the opercular side by fan organs, which are modified dorsal filaments (Figure 4.10). The inhalant opening is also protected by a fan organ in *Anabas,* but not in all anabantoids. During water ventilation the fan organs will seal the openings of the air sac under hydrostatic pressure. There may also be passive opening by ventral movements of the fan organs during air ventilation, in which case it is envisaged that deep buccal-floor movements may bring about fan-organ opening and closing via coupling through the gill arches.

Air ventilation is not monophasic in all anabantoids. Juvenile *Anabas* and a gourami, *Osphronemus goramy,* ventilate the air sac in a bidirectional (diphasic) manner (Peters 1978). *Helostoma temmincki*, interestingly, employs both monophasic and diphasic modes of ventilation. Diphasic ventilation differs from the monophasic mode in that exhalation through the mouth occurs passively under hydrostatic pressure

Figure 4.8. Simultaneous buccal pressure (solid line) and air bladder pressure (broken line) during a single air bladder ventilation in jeju. Inhalation (I) precedes exhalation (E) (after Farrell and Randall 1978). (Reproduced by permission of the National Research Council of Canada from the *Canadian Journal of Zoology* 56: 939–45, 1978.)

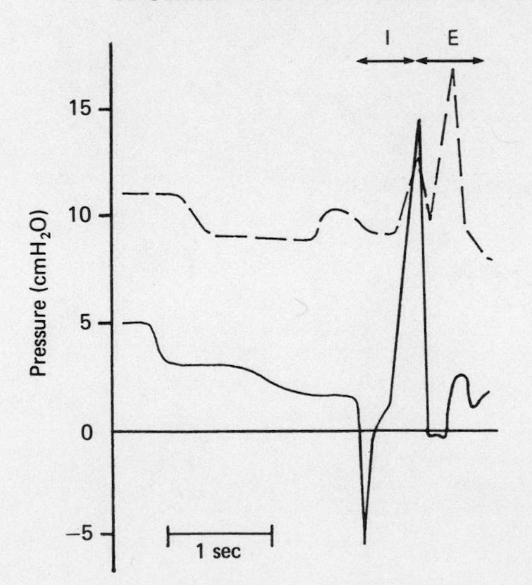

and is retrograde to inhalation. Water is allowed to enter the suprabranchial chamber through the opercular openings and to displace air into the buccal cavity and out through the mouth. Subsequent air inhalation with the buccal pump displaces the water from the air chamber. Adult *Anabas* have selected the rapid monophasic ventilation which excludes water from the air chamber and is seven times faster (0.15 sec vs. 1.1 sec) than diphasic ventilation found in the juvenile.

Buccal breathing

Probably the simplest modifications for air breathing in fish are the highly vascularized buccal and/or pharyngeal cavities (e.g., electric eel, *Electrophorus electricus*) or the opercular cavity (e.g., *Symbranchus marmoratus*). The gills may also be structurally modified to prevent lamellar collapse during air breathing (e.g., *Gillichthys mirabilis*; Todd and Ebeling 1966). These fish simply "bite off air" and retain it within

Figure 4.9. Breathing pattern of a lungfish (1.8 kg) that had been aestivating in a cloth sack for 3 months. The tachypneic period is divided into two phases: Phase I consists of rapid shallow breathing; phase II starts at the end of the tachypnea after the last expiration and consists of a series of large inflations that fill lungs to preapneic levels. A, decrease in buccal pressure at start of lung ventilation period; B, movement of air into mouth; C, first emptying of lungs (downward deflection in mouth opening air flow trace), probably due to an initial active component to expiration; D, air movement into mouth at start of phase II; E, posterior buccal pressure increase, which pumps air into lungs; F, simultaneous increase in lung pressure associated with filling of lungs by buccal force pump; G, decrease in buccal pressure similar to A; H, final inflation of lung by large increase in posterior buccal pressure; I, posttachypneic period rise in lung pressure during apnea (from DeLaney and Fishman 1977).

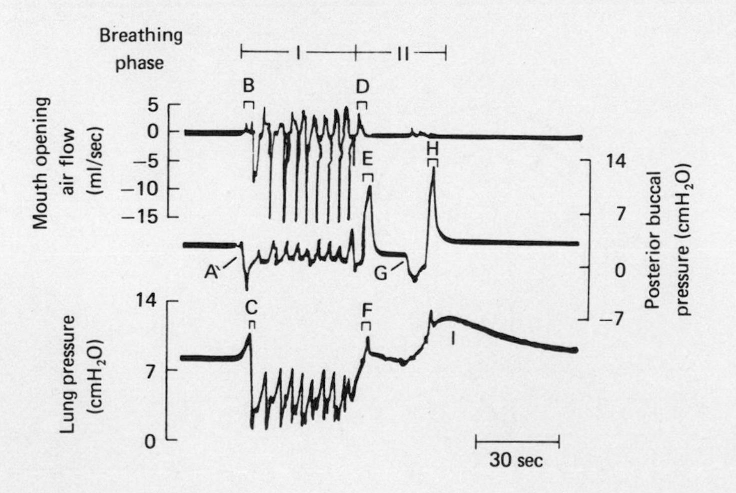

their buccal or opercular cavity before releasing it via their mouth. Air is drawn by suction into the air-breathing organ, and ventilation is tidal.

Patterns of air flow during ventilation

From the previous outline of ventilation mechanisms it is clear that several different modes of ventilation exist. The most common mode in

Figure 4.10. *Osphronemus goramy*, structure of accessory breathing organs in fish of about 40–45 mm total length (*schema*). (*a*) Total view, suprabranchial chamber opened by incision through the wall. (*b*) Part of (*a*) without opercular apparatus, (at) atrium; (dc) caudodorsal compartment; (tro) trough; (ip) inner plate; (dp) double plate; (cp) border of the closure plate; (pho) pharyngeal opening; (sh) shutter on first gill arch (I); (st) stylet; (bo) branchial opening between first and second gill arch (II). Lamellae of first gill arch in this region removed (after Peters 1978).

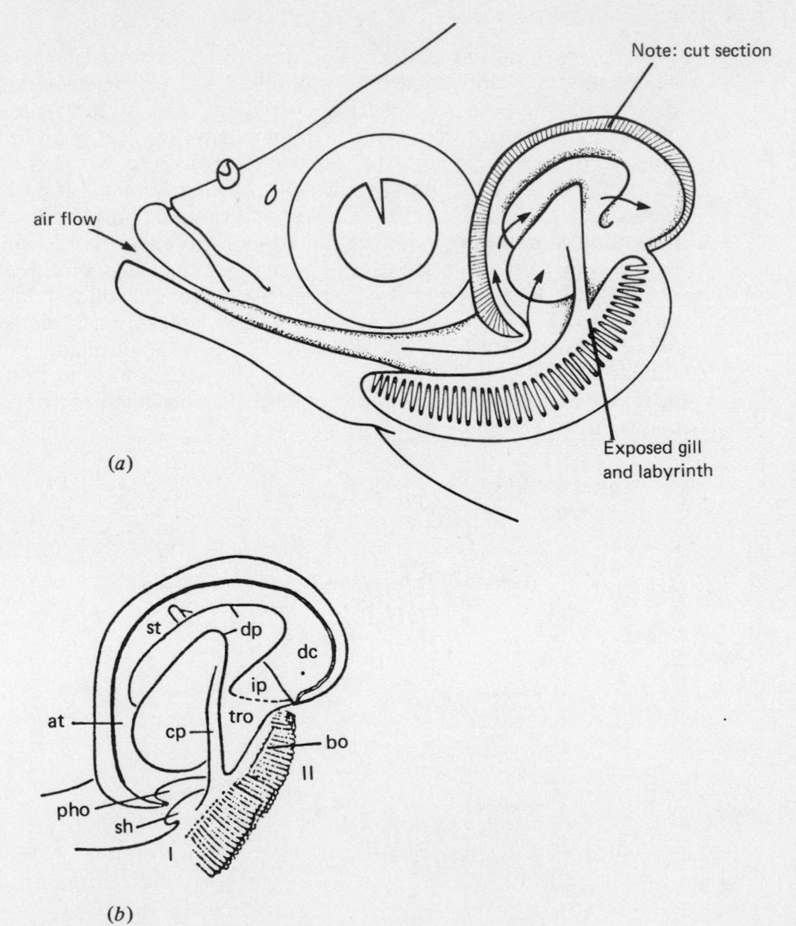

air-breathing fish is a biphasic pattern, where expiration precedes inspiration.

A sequence of "inhalation then exhalation," however, is found in some air-breathing fish, for example, *Hoplerythrinus* (Farrell and Randall 1978), *Piabucina* (Graham et al. 1977), and probably *Erythrinus*. This sequence is primarily a consequence of the danger from aerial predation to these fish when they surface. The advantage of this sequence is that surface exposure can be terminated any time after the first inspiration is completed (i.e., 0.2 sec). However, the sequence has two possible drawbacks: (1) mixing of inhalant and exhalant gases in the buccal cavity and/or bladder; and (2) overinflation of the gas bladder. A structural adaptation of the gas bladder alleviates both problems. The gas bladder is divided into two distinct anatomical regions by a muscular sphincter: a highly compliant but muscular anterior nonrespiratory chamber and a relatively stiff, posterior respiratory chamber (Kramer 1978). Inhaled gas flows preferentially into the anterior chamber, and "jet streaming" reduces mixing with residual gas in the posterior chamber of the bladder (Figure 4.11a). The normally collapsed anterior chamber expands to prevent overinflation of the almost full posterior chamber. Exhalation from the posterior chamber occurs before the inhalant air is released from the anterior chamber into the posterior one.

Jet streaming is probably important in the triphasic ventilation pattern of lungfish and anurans. In these animals, part of the buccal inspiratory cycle precedes exhalation. The possible mixing of inhalant and exhalant gases in the buccal cavity is avoided by jet streaming (Figure 4.11b). This mode of ventilation is probably designed to prevent water entry into the lung because the likelihood of drowning is remote if the buccal cavity is cleared of water before the glottis opens.

Selective advantages and disadvantages of buccal pump ventilation

Amphibians and some air-breathing fish have a modified buccal pump for air breathing. The large dimensions of the jaw in Amphibia give a mechanical advantage because buccal volume determines the power output of the buccal pump, at least in *Rana pipiens* (West and Jones 1975b). Also a large buccal volume means a large tidal volume, and therefore a higher ventilation volume. These vertebrates are, however, predominantly aquatic air breathers, suggesting that modified buccal pump mechanisms have had limited evolutionary scope on land. Indeed, the jaw is necessarily light in structure, such that diet becomes restricted to vegetation, insects, and other small animals. Also, the structural limitations on buccal size imposed by feeding determine tidal volume and the efficiency of lung ventilation. Aspiration breathing does

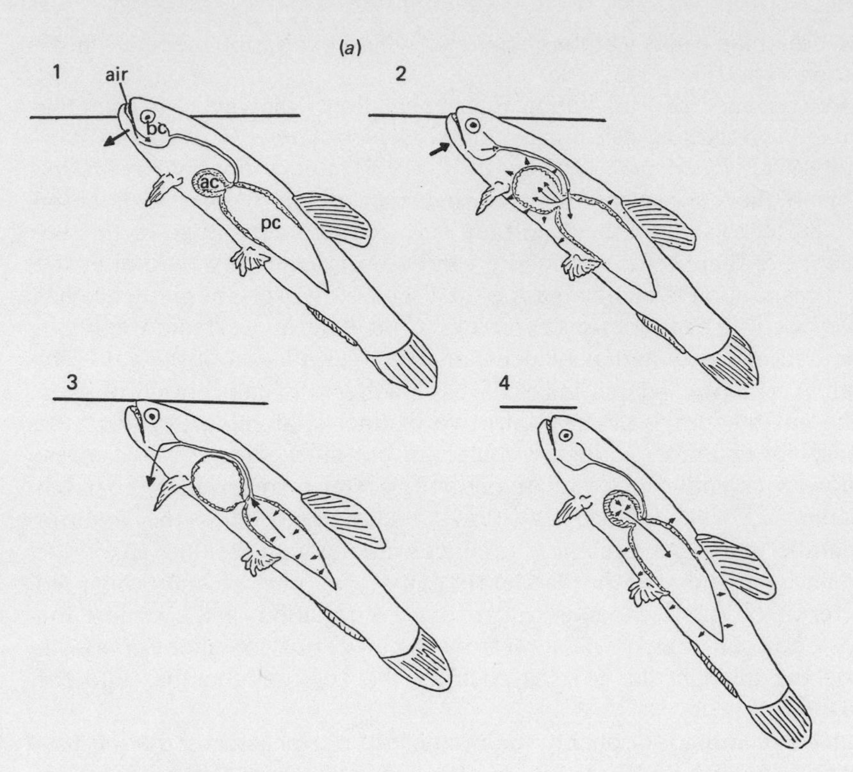
(a)
1
air
bc
ac
pc
2
3
4

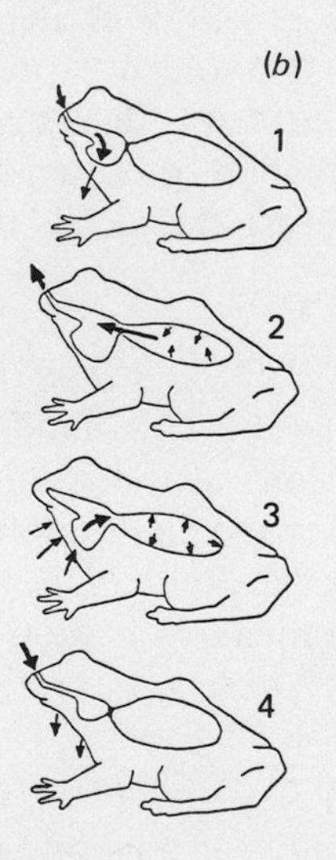
(b)
1
2
3
4

not impose the above restrictions but allows modification of the mouth and jaws for feeding.

Evolution of aspiratory ventilation

The divorce of buccal structures from lung ventilation undoubtedly had an important evolutionary impact. Reptiles are representatives of terrestrial air breathers using aspiration. They are able to fill a different terrestrial niche from that of amphibians as a direct consequence of their different ventilatory mechanisms. Reptiles are predatory and possess small heads, strong jaws, and armored skin, important for that role.

Aspiration breathing probably evolved in an aquatic form that also possessed a buccal pump, and acted to supplement bladder ventilation. The buccal pump could still be used for both bladder and gill ventilation, and the two mechanisms are not mutually exclusive. As long as gills are retained, so will a buccal pump be, even if the bladder or lung is ventilated by aspiration. In fact terrestrial snakes and tortoises have retained a buccal pump for sampling air through their nostrils. The best example we have of the simultaneous use of aspiration and buccal pump ventilation is in the obligate air-breathing fish, the pirarucu, *Arapaima gigas*. This fish ventilates its gas bladder by aspiration (Figure 4.12) and its gills with a buccal pump. As a consequence, the tidal volume exceeds buccal volume, and the lung volume is high, 10% of body volume (Farrell and Randall 1978). The gills are ventilated by a buccal force pump in a typical teleost fashion, but opercular move-

Figure 4.11. Jet streaming during air breathing in fish and frogs. (*a*) *Hoplerythrinus*. (1) Air is drawn into buccal cavity (bc). (2) Air flows preferentially into the muscular anterior chamber (ac) by jet streaming, with buccal cavity compression. The pressure will also rise within the posterior chamber (pc). (3) Air is exhaled from pc with ac closed off. (4) Air is released from ac to pc to complete the ventilation cycle. There may be further exhalation during this period. (*b*) The frog, *Rana catesbeiana*. (1) With the glottis closed and nostrils open, the buccal floor drops owing to gravitational and elastic forces, aspirating air, particularly into the postero-ventral portion of the buccal cavity. (2) The glottis opens, and air is driven along the dorsal roof of the buccal cavity and out through the open nostrils by the inherent elasticity of the lung and body wall. (3) The nostrils close, and contraction of the buccal and particularly the petrohyoid musculature forces the gas stored in the postero-ventral portion into the lung, distending its walls. (4) The glottis closes, and the nostrils open again, after which gravitational and elastic forces cause the buccal floor to drop, aspirating air into the buccal cavity. A series of oscillatory buccal contractions with open nostrils next flush out the buccal contents prior to the following pulmonary ventilation cycle. (Part *b* from Gans 1970.)

ments are minimal. The buccal pump also contributes to gas-bladder filling. The mechanics behind aspiration ventilation in pirarucu are unclear. The gas bladder is highly modified and is no longer saclike; instead a nonmuscular septum extends between the highly vascularized flank regions. We suggest that reduced pressures are generated in the lung by arching of the body as the fish surfaces for an air breath. This would cause lateral expansion of the heavily ribbed flanks, which is translated to a downward stretching of the septum (Figure 4.13).

The variety of reptilian modes of lung ventilation differ only in detail with respect to the mammalian aspiratory mode of ventilation. They, too, are characterized by subatmospheric intrapulmonary pressure during inhalation, which is generated by movements of either the ribs or the diaphragm. The differences between lung ventilation in reptilian

Figure 4.12. Aspiratory ventilation in *Arapaima*. (*a*) Air bladder pressure during two air breaths. (*b*) Detailed, simultaneous pressure recordings from the buccal cavity (solid line) and air bladder (broken line) during an air breath. Note that the air bladder pressure is lower than buccal pressure during inspiration (I), which follows expiration (E) (from Farrell and Randall 1978). (Reproduced by permission of the National Research Council of Canada from the *Canadian Journal of Zoology* 56: 939–45, 1978)

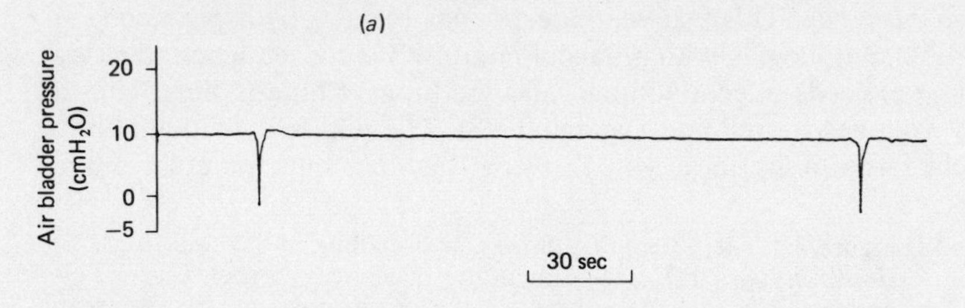

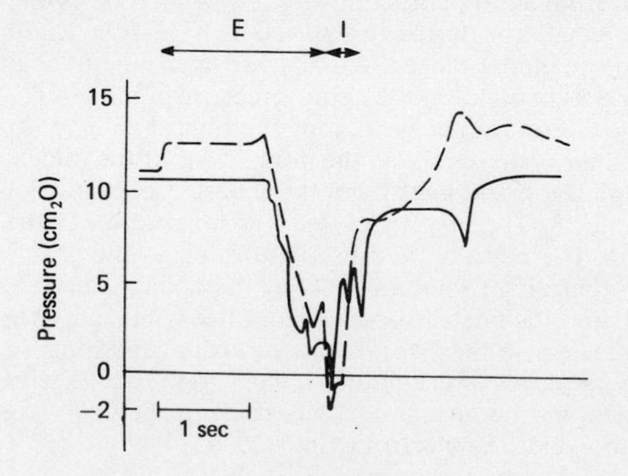

subclasses are a result of compromises between ventilation mechanisms, the rigidity of the body walls, and the environmental niche of the reptile in question. A brief outline of ventilation in four reptilian subclasses highlights the variety of terrestrial aspiration mechanisms and the modifications necessary for aspiration breathing in the aquatic environment.

Breathing without the use of ribs

The Chelonia aspirate their lungs through movements of the diaphragm (Gans and Hughes 1967). The diaphragm is a nonmuscular septum onto which the viscera are attached, separating the pleura-enclosed lungs from the body cavity. During inhalation in the tortoise, *Testudo graeca*, the diaphragm falls with changes in body volume when the pectoral and pelvic limbs are extended out of the shell (Gans and Hughes 1967). Exhalation usually precedes inhalation and is produced by limb retraction. The increased pressure in the lung maintains exhalation even after limb retraction stops (Figure 4.14). The problem with these mechanisms is that any change in body volume, as occurs during walking or neck extension, will be superimposed on the intrapulmonary pressure. Thus lung ventilation is regulated by active glottal opening and closing. In aquatic species (e.g., snapping turtle, *Chelydra serpentina*), the body is supported by hydrostatic pressure. When the animal is fully submerged, expiration is passive and due to hydrostatic pressure, and pelvic girdle movements increase lung volume during inspiration. On land the reduced plastron cannot support the visceral weight, and the

Figure 4.13. A diagrammatic cross section through the body of *Arapaima* in the region of the highly developed air bladder, which is separated from the viscera by a septum. Lateral flank movements as the fish surfaces and arches its back cause the septum to move in a downward fashion to draw air into the air bladder.

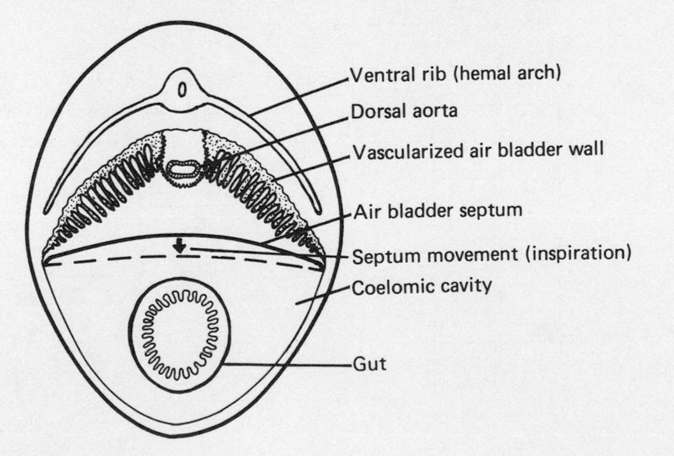

diaphragm is depressed, with production of below-ambient intrapulmonary pressures (Figure 4.14). Thus inspiration is largely passive, and expiration is achieved through pelvic girdle retraction (Gaunt and Gans 1969). At intermediate levels of carapacial submergence there is a gradual transition between active expiration and active inspiration (Figure 4.15).

Breathing without the use of a diaphragm

Crocodilians possess well-developed ribs and intercostal muscles, but lack a diaphragm. Aspiration is achieved by contraction of the m. diaphragmaticus, attached to the liver and pelvic girdle, which draws the liver posteriorly to expand the lungs. Contraction of various abdominal muscles posterior to the liver returns the liver forward during expiration (Figure 4.16) (Gans and Clark 1976). The intercostals appear to maintain ribcage structure rigid during ventilation rather than alter

Figure 4.14. Diagram comparing the three pressure phases of the ventilatory cycle for *Testudo* (left) and *Chelydra* (right). The broken line in the *Chelydra* pressure record gives the pressure curve when the shell is flooded, the solid line the curve when the animal is on land. Note that *Chelydra* shows more irregularities in muscle activity, which correlate with the amount of water in which the turtle is immersed. (Left side after Gans and Hughes 1967; right side after Gaunt and Gans 1969.)

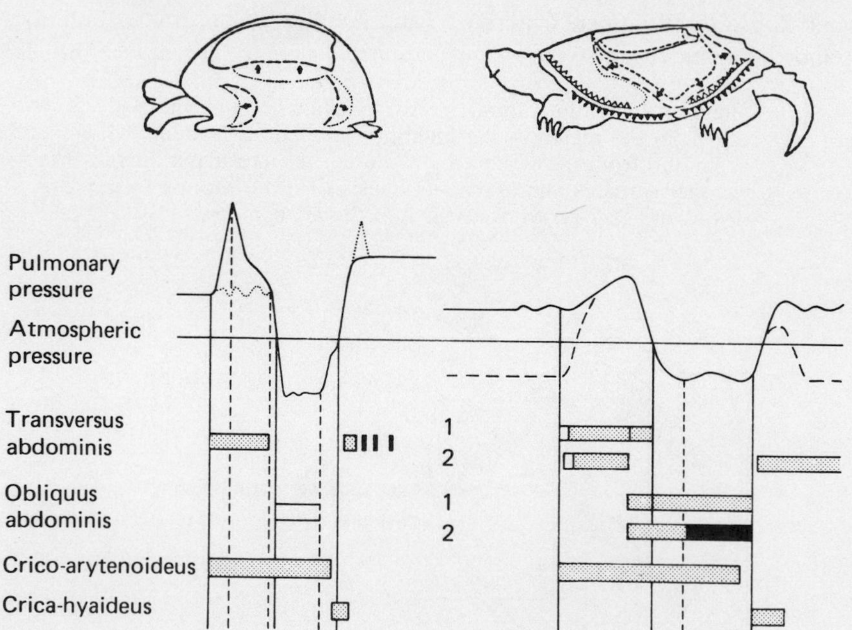

thoracic volume per se. The piston-type movements of the liver are ideally suited for heavy, short-limbed tetrapods that rest on their ventrum, making ribcage movements difficult and restricted. In lizards, which have well-developed legs compared to crocodiles, the intercostals actually power inhalation and exhalation. In water, crocodiles utilize hydrostatic pressure to move the liver anteriorly during expiration.

Breathing without the use of legs and/or diaphragm

Snakes lack a diaphragm and well-defined pleura, but the rib development and its supporting musculature are extensive. The lung extends for much of the body length, but is only partially alveolated (20 to 50% depending on the species; McDonald 1959), and the saccular lung is nonrespiratory (Figure 4.17). The intercostal muscles generate regional thoracic expansion for lung aspiration, and compression for exhalation (Figure 4.17). There are also passive components to inspiration and expiration. Resting tidal volume is low, and only a small but variable region of the body wall is involved in lung ventilation at any time. Full participation of larger regions of the body wall results in much higher tidal volumes. Any area of ribs can bring about ventilation provided the ribs surround a portion of the lung, even a saccular portion. Clearly, this aspiratory mechanism has tremendous adaptive value for the snake. Ventilation can proceed while the intercostals in some regions are otherwise engaged or incapacitated by other activities, such as locomotion, digestion of large prey, coiling, or constricting. Aspiration breathing in aquatic sea snakes has not been examined.

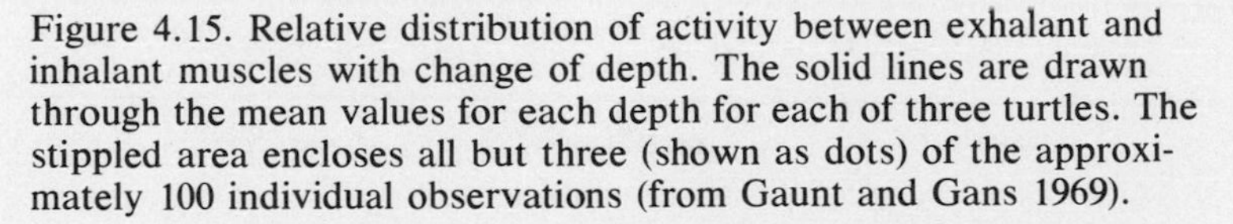
Figure 4.15. Relative distribution of activity between exhalant and inhalant muscles with change of depth. The solid lines are drawn through the mean values for each depth for each of three turtles. The stippled area encloses all but three (shown as dots) of the approximately 100 individual observations (from Gaunt and Gans 1969).

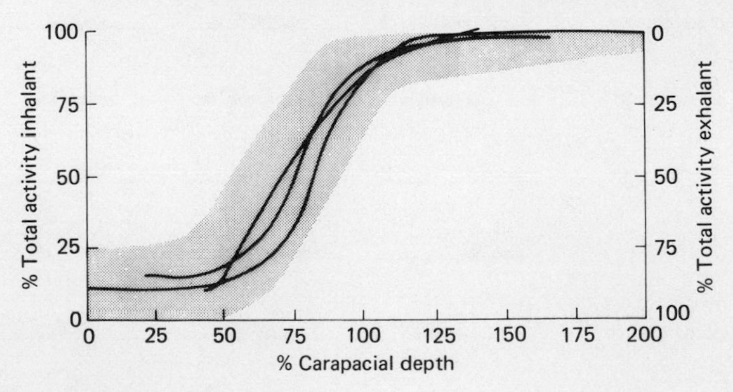

Figure 4.16. *Caiman*. (*a*) Sketch to show the major structures and the muscles tending to influence filling and emptying of the lung. (*b*) Activity patterns of various muscle groupings at different depths of immersion, combined from a series of observations of animals over a four-month period. The activity recorded from any intercostal layer depends on the site of electrode implantation. (After Gans and Clark 1976.)

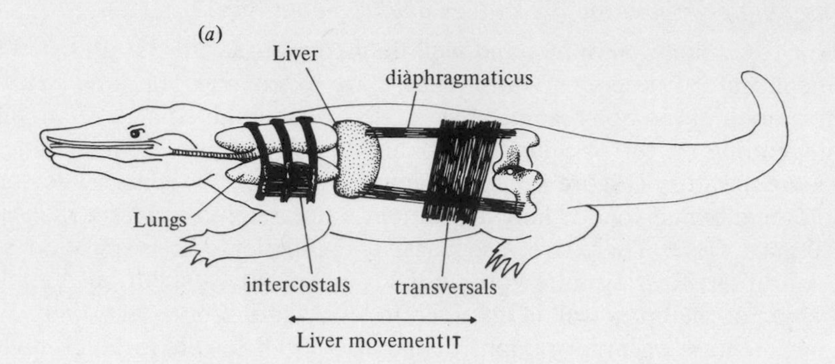

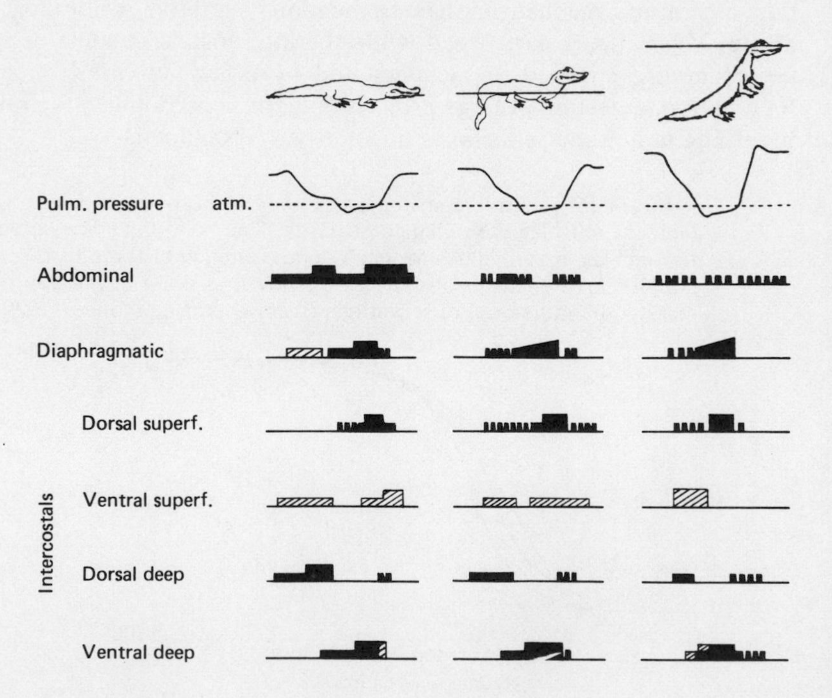

Patterns of reptilian ventilation

Patterns of reptilian ventilation are characteristically varied, but two modes are common. A series of ventilations, each interspaced by a few seconds, is separated by much longer periods of apnea, 1 min to 1 hr, or single ventilations are separated by an apnea lasting on the average 0.2 to 2 min (Figure 3.3). The expiration–inspiration cycle is on the whole of long duration in comparison with other vertebrates. Snakes, for in-

Figure 4.17. The mechanism of lung ventilation in snakes. (atm) atmospheric pressure; (dl) dorsolateral; (vl) ventrolateral (adapted from Rosenberg 1973 and Macintyre and Farrell unpublished data). The sketch of the snake shows the alveolar (respiratory) and saccular (nonrespiratory) portions of the lungs.

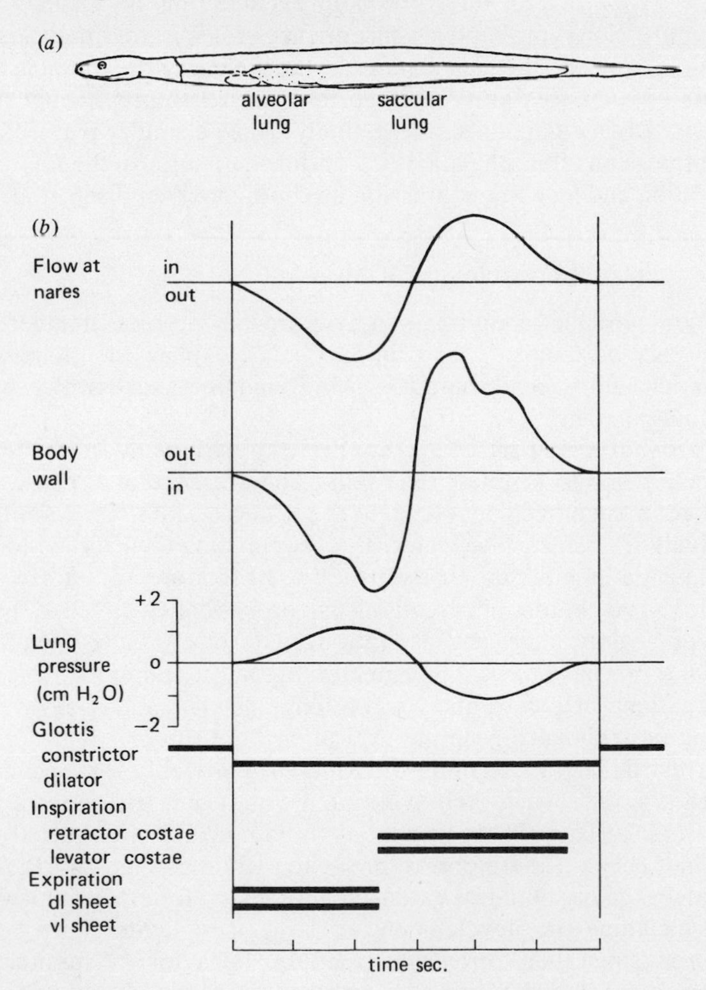

stance, inflate and deflate their lungs over a period lasting 10–20 sec, and, if rushed and excited, take a mere 5 sec (Rosenburg 1973; Clark et al. 1978). *Chrysemys* typically takes 3–4 sec for completion of the air flow cycle.

In quiescent reptiles the air flow cycle is biphasic, expiration preceding inspiration, despite triphasic and often quadraphasic intrapulmonary pressure oscillations (Gans 1976). Triphasic air flow has been reported in some reptiles in certain situations, consisting of a second, small expiration following the inspiration during lung recompression. Triphasic flow, usually found in excited animals, is not always correlated with triphasic intrapulmonary pressure (Clark et al. 1978), and a biphasic flow pattern has recently been advanced as the general reptilian condition (Gans and Clark 1978).

After inspiration the lung is recompressed during the whole of apnea. Recompression is probably a consequence of inspiratory muscles relaxing, thus exposing the closed lungs to the weight of the thoracic cavity. Expiratory muscles may be important for recompression in aquatic reptiles. Diving mammals, interestingly, show a similar pattern of lung recompression after inhalation (i.e., prior to diving). In the Pilot whale, exhalation and lung recompression are both passive (Olsen et al. 1969).

The evolution of breathing mechanisms

It is now possible to envisage an evolutionary pathway for vertebrate ventilatory mechanisms, and, in particular, explain the dichotomy of aspiration and buccal pump breathing, and the success of aspiration breathing on land.

In pre-Devonian periods all vertebrates were aquatic, probably using a buccal pump to ventilate their gills. The presence of hypoxic waters resulted in the movement of fish to the water–air interface to utilize the relatively O_2-rich surface water. Buccal pumping then served to inflate the air-breathing organs that were evolved to store and utilize atmospheric O_2. Aspiratory air-breathing fish probably developed at the same time or slightly later, and aspiration could easily have been supplemental to, or have been supplemented by, buccal pumping. Aspiratory air-breathing fish were likely to be long, slender, and heavily ribbed, relying on arching to generate aspiration.

Terrestrial migrations of air-breathing fish probably first became prolonged at a time when there were plants and some arthropods, but no predators, on land. These ancient air-breathing fish with buccal pumps were not subject, therefore, to predation while on land. Furthermore, the buccal cavity could be extensively modified for efficient air ventilation, including the development of larger and lighter jaws, without compromising their browsing feeding behavior. Aspiratory air-breathing fish faced a different problem during terrestrial migrations.

They required limb support greater than that which could be afforded by strengthened pectoral fins. They could maintain a predatory behavior, however, because they had heavy jaws and armored skin. Air-breathing fish with a buccal pump were clearly faced with the simplest modifications, and probably evolved onto land first. Since there was no terrestrial predation, they flourished.

In the early Amphibia, CO_2 excretion was switched from the gills to a moist skin, and the gills and scales were lost (Chapter 2). The problem of dehydration, however, tied early amphibians to water even though limb development (tetrapody) extended the range of terrestrial feeding. Such is still the case in modern anurans. Aspiration air breathers also evolved tetrapody, allowing them to assume a terrestrial predatory role. Upon what did they prey? It is likely that their main prey were browsing buccal pump breathers (early amphibians, Dipnoi, and possibly other fish on land) as well as other aspiratory breathers. To account for the abundance of reptiles during the early Mesozoic era, their potential terrestrial prey, the amphibian stock, had first to flourish and establish its niche.

The first aspiratory-breathing vertebrates were probably chained to water for CO_2 excretion and pH regulation. Some, like the temnospondyl, *Eryops*, the most common of early Permian amphibious forms (Romer 1972), may have retained gills for CO_2 excretion, had a scaly skin, and ventilated their lungs by aspiration. The excretion of CO_2 via the lung would essentially uncouple these animals from the requirements of gill ventilation with water. Animals such as these are probably typical of the ancestral reptilian stock. Reptiles are generally predators and have a scaly skin. They excrete CO_2 via their lung, and regulate CO_2 levels and pH via lung ventilation. Tidal aspiratory ventilation is well suited for controlled CO_2 excretion. A fine control of lung CO_2, and therefore blood pH, can be achieved by precise alterations in ventilation through tidal volume, as well as frequency.

More efficient predatory traits were selected for (e.g. faster locomotion, reduced weight, and high ventilation rates to meet the requirement of a higher metabolism). Lung volume could be increased without compromising the limits set on buccal size in buccal pump air breathers. Furthermore, higher ventilation volumes were achieved by increasing tidal volume as well as breathing frequency. Mammals are presently successful terrestrial species: predators possessing good locomotion, rapid ventilatory acid–base control, and aspiration breathing.

We suggest that both buccal and aspiration air ventilation evolved in the aquatic environment. The early air-breathing vertebrates contained two lineages: one with buccal pump ventilation giving rise to modern-day anurans, the other lineage with aspiratory mechanisms giving rise to reptiles, birds, and mammals.

5 *Regulation and control of gas transfer*

Common to all terrestrial air breathers and their bimodal and aquatic antecedents is the need to achieve a balance between O_2 consumption–CO_2 excretion in the tissues and the rate of delivery or removal of these gases via the gills, bladder, skin, or lungs. This balance between "supply" and "demand" normally is very finely controlled in some groups of vertebrates, whereas in others it may be achieved only in the relatively long term.

Continuous water breathers

Almost all water breathers respond to hypoxic exposure with an increase in the frequency and amplitude of branchial pumping (see Shelton 1970b for review). The extraction of O_2 from the water in aquatic breathers is usually low, but the oxygen stores of the body, intervening between the respiratory membranes and the mitochondria, are limited to comparatively small amounts bound to hemoglobin and myoglobin. Thus, overall there is not a high oxygen-store-to-oxygen-utilization ratio (Chapter 3), a situation that severely restricts the short-term ability of aquatic fish to protect the tissues from the effects of environmental hypoxia or changes in metabolic rate. This low store-to-utilization ratio has necessitated a sensitive control over both ventilation and perfusion to maintain oxygen supplies to the tissues. Such a control system has to (1) provide monitoring of respiratory gas levels, particularly O_2, at one or more sites between the respiratory membranes and the mitochondria, and (2) mediate active behavioral (avoidance) and physiological (hyperventilation, lamellar recruitment) responses, such that even brief imbalances in O_2 supply and demand are obviated. Even in those few unusual bottom-dwelling fish such as juvenile *Ictalurus* and *Acipenser*, which lower gill ventilation and oxygen uptake upon hypoxic exposure (Gerald and Cech 1970; Burggren and Randall 1978), these responses indicate that they are detecting and responding in their own way to changes in environmental oxygen.

Whether or not an aquatic fish increases gill ventilation during hypoxic exposure, it must also effect an appropriate cardiovascular response to optimize the ventilation : perfusion ratio of the gills, and thus optimize branchial gas exchange. This may not require simply an increase in perfusion to maintain a constant ventilation : perfusion ratio. Generally fish respond to hypoxia by reducing heart rate but increasing stroke volume, such that cardiac output is either maintained or increased (Randall 1970a). Lamellar recruitment occurs and so, even though cardiac output is somewhat elevated, transit time for blood through a single lamella is increased (Farrell 1979). Blood is probably redistributed within secondary lamellae, resulting in a reduction in diffusion distance between blood and water (Farrell 1979). Thus, in most water-brething fish, an increase in transit time for blood flow through lamellae, lamellar recruitment, reduced diffusion distances, and increased gill water flow all act in concert to maintain oxygen uptake during hypoxia. Similar changes, mediated by different mechanisms, probably occur during exercise, except that there is a marked increase in cardiac output owing to an increase in both heart rate and stroke volume.

Fish respond to increases in CO_2 by increasing gill ventilation (Janssen and Randall 1975), but this response can be either ameliorated or completely abolished by high oxygen (Randall and Jones 1973), and it is possible that CO_2 effects on gill ventilation can be explained by secondary effects that elevated CO_2 levels have on oxygen transport in the blood. The regulation of gill ventilation is related to oxygen delivery rather than CO_2 removal (Randall and Cameron 1973), for during hypoxia the increased gill ventilation causes a respiratory alkalosis (Figure 5.1), and hyperoxia, which results in a marked reduction in gill ventilation, causes a marked respiratory acidosis due to CO_2 retention. Even after 10 days of hyperoxic exposure there is no tendency to readjust gill ventilation upward to restore the carp's normal acid–base status, as long as O_2 uptake can be met with a reduced ventilation. In view of these unregulated swings in blood CO_2 and pH, it is not surprising that in water breathers, generally, ventilatory and cardiovascular responses to increasing environmental CO_2 are much attenuated and sometimes absent. The primary adaptation to hypercapnic exposure in water breathers is an increase in the bicarbonate content of the body fluids to reduce pH fluctuations (Randall and Cameron 1973).

While the physiological responses of water breathers to changes in environmental and/or internal gas partial pressures have been well documented, the afferent side of the ventilatory and cardiovascular reflexes is still largely an enigma. Chemoreceptors responsive to changes in environmental gases could be located in the pharynx, on the surface of the gills, or in the efferent branchial circulation, whereas changes in blood gases could be monitored in the efferent or in the afferent branchial circulation. On first appearance the greatest potential for maintaining a homeostasis under such conditions, by altering both ventilation and perfusion, would arise from a control system detecting gas partial pressures in the bloodstream, and particularly in the efferent branchial circulation (Jones et al. 1970). Here, all of the multitude of factors influencing oxygen diffusion across the gills (see Chapter 2) become manifest in one readily detectable variable – arterial blood P_{O_2}. A control system simply dependent on this single afferent pathway, however, would be unable to differentiate between a potential fall in arterial P_{O_2} caused, say, by a reduction in inspired P_{O_2}, and that caused by the increased metabolic rate, tissue oxygen extraction, and cardiac output that would result from exercise. From a biochemical point of view this is immaterial, as O_2 delivery to the tissues can be maintained in either instance, but from a physiological point of view the differentiation is important because different responses must occur to maintain O_2 delivery. Additionally, such a limited chemoreceptor complement robs the fish of important information on the quality of the environment, according to which it can modify its behavior. Chemoreceptors reflexly adjusting ventilation and/or perfusion in water breathers have been reported to reside variously in the pseudobranch, the first gill arch, the dorsal aorta, and the brain (Elancher 1974; Elancher and Dejours 1974; Bamford 1974; Smith and Jones 1978; Daxboeck and Holeton 1978). We believe it highly likely that multiple receptor sites for oxygen sensitiv-

Figure 5.1. Bicarbonate concentration [HCO_3^-] and partial pressure of CO_2, P_{CO_2}, and pH of arterial blood sampled in the dorsal aorta via an indwelling catheter in a large carp at 15°C. Abscissa, time in days. The partial pressure of O_2 in the water, P_{WO_2}, is indicated at the top of the figure (from Dejours 1973).

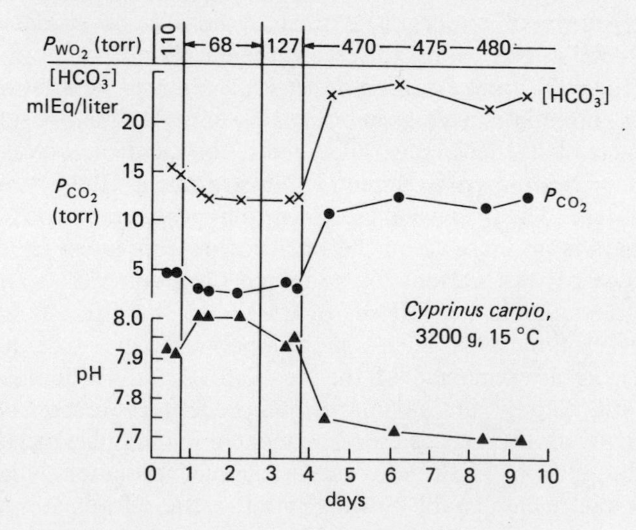

ity, both external and central, must exist in every water-breathing vertebrate in order to account for the wide range of cardiovascular, ventilatory, and behavioral responses observed. Additionally, in aquatic fish there are mechanoreceptors lying in the inhalant water stream, and although rhythmic ventilatory movements still occur in the absence of afferent mechanoreceptor activity, they seem to play a role in the regulation of gill ventilation. Thus, chemical and mechanical afferent information is available to a central integration in the brain stem, which is then responsible for the appropriate effector response (Shelton 1970b).

It is not clear to what extent ventilation and perfusion of the gills are independently controlled in aquatic breathers. For example, during hypoxic exposure gill ventilation increases and heart rate decreases (Randall 1970b), whereas during exercise both variables increase (Jones and Randall 1978). Under certain, often stressful, circumstances there is a close coupling of the phasing of ventilation and perfusion in water-breathing fish (Satchell 1961; Hanson 1967; Randall and Smith 1967), which would indicate interactions at the central as well as the peripheral level. Whatever the as yet undescribed intricacies of the control system for gas exchange in continuous water breathers, they must respond nearly instantaneously, for there are effectively no oxygen stores to buffer the tissues while awaiting readjustments in ventilation and perfusion.

Intermittent and bimodal breathers

Every bimodal breather perfuses, either in series or in parallel, at least two distinct gas exchange organs, which in turn are ventilated, sometimes intermittently, by two quite different respiratory media. Moreover, the various respiratory organs will contribute disproportionately to oxygen and to carbon dioxide exchange. Bimodal, intermittent breathers usually have a large O_2 store resulting from the incorporation of a gas-filled bladder or lung, as well as the ability to buffer and store large amounts of CO_2 in various body-fluid compartments. The metabolic rate, and hence rate of oxygen extraction from the O_2 stores, is low, however; so the ratio of oxygen store to oxygen extraction is quite high, distinguishing these animals from continuous water or air breathers. As we have indicated in Chapter 3, this circumstance allows most bimodal breathers to ventilate their air-breathing organs intermittently because the tissues are quite well protected from transient interruptions in oxygen uptake by the presence of a large oxygen store.

Most air-breathing fish and other bimodal breathers detect and respond to, often with similar sensitivity, both hypercapnia and hypoxia in the inspired water or air (Johansen 1970). The extent of the changes in perfusion and ventilation of a particular organ, however, depends on many factors, including the importance of that organ in the exchange of either O_2 or CO_2.

There are thus many varied ventilatory responses to a changing environment that can occur in bimodal breathers (Figure 5.2). The usually complex arrangement, in parallel or in series, of the circulation of the aerial exchange organ and the gills necessitates carefully coordinated cardiovascular adjustments to ensure optimal ventilation–perfusion matching and thus optimal gas exchange (see Figure 3.1). In *Electrophorus*, for example, hypoxia in inspired gas produces not only a hyperventilation of the buccal cavity, but also a tachycardia, plus a considerable redistribution of cardiac output to the mouth and away from the body tissue (Figure 3.7). Clearly, a much more elaborate control system for ventilation, perfusion, and their coordination than that found in any of the continuous water breathers has had to evolve along with the air-breathing life style. This is particularly true because most bimodal breathers live in a habitat where the levels of respiratory

Figure 5.2. Relationship between the type of respiration and the oxygen and carbon dioxide content of water in *Erythrinus* (from Willmer 1934). (dotted circles) aquatic respiration, (filled circles) intermediate respiration; (+) aerial respiration.

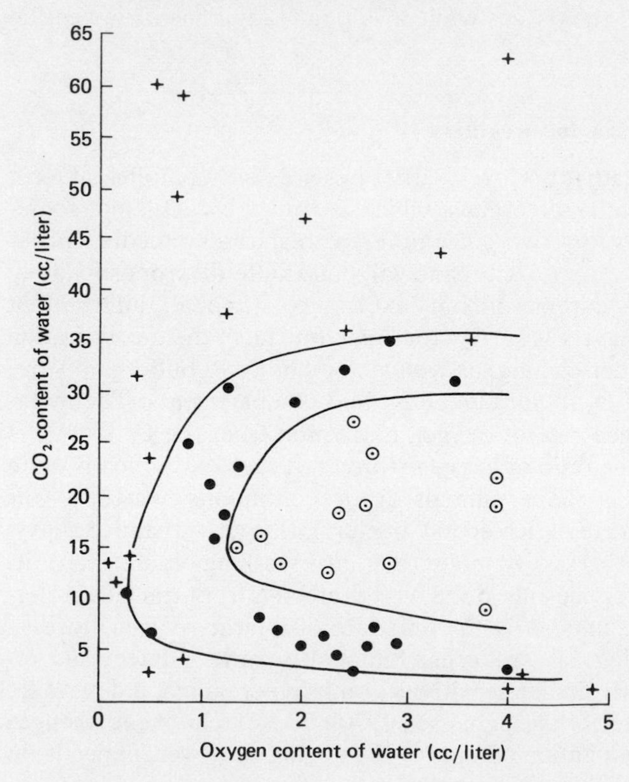

gases in water may often fluctuate widely and rapidly – after all, we presume this to be a major selection pressure for air breathing in the first place.

Air-breathing fish and other bimodal, intermittent breathers appear to retain the chemoreceptor complement evident in the water breathers, that is, receptors located both in the systemic arterial bloodstream (Singh and Hughes 1973) and externally to monitor the quality of inspired water (Johansen 1966; Johansen and Lenfant 1968). A few air-breathing fish (e.g., *Trichogaster*) appear to be unable to differentiate between hypoxia in the inspired gas and inspired water (Burggren 1979). A hypoxic atmosphere may rarely, if ever, be encountered by the gourami in nature; so normally a hyperventilation of the suprabranchial chamber stimulated by low blood oxygen levels would be the appropriate response to increase blood oxygenation. Thus, there may have been no selection pressures acting on such air breathers to evolve a chemoreceptor system capable of discriminating between the oxygen (and carbon dioxide) levels of the different respiratory media ventilating the gills and the air-breathing organ.

In other bimodal breathers, particularly those that are much more dependent on aerial gas exchange, information on the performance of the aerial exchanger itself, distinct from other exchangers, becomes increasingly important. Perfusion – and thus O_2 depletion and CO_2 addition – of the lung or bladder during apnea must be closely regulated, so that these aerial exchangers can function as a long-term oxygen store and CO_2 sink during apnea. The degree of O_2 depletion and CO_2 build-up at any given time could be regulated by chemosensitive receptors monitoring bladder or lung gas, or efferent blood from these organs. The existence in air-breathing fish of bladder or lung receptors as antecedents of those in the continuous air breathers is equivocal. However, chemosensitive elements have been identified in the lung parenchyma of amphibians (Rogers and Haller 1978) and reptiles (Milsom and Jones 1976).

Alternatively, arterial blood chemoreceptors could mediate appropriate ventilatory and cardiovascular responses to terminate apnea. A clear relationship exists between arterial blood P_{O_2} and the diving duration in the turtle *Chelys fimbriata* (Figure 5.3), suggesting that a blood oxygen stimulus of some sort is responsible for terminating, and perhaps initiating, apnea. In these animals arterial blood partial pressures reflect not only lung performance, but also the extent of the central intracardiac blood shunt. Hence, the monitoring of arterial blood gas partial pressures provides useful information only if, in addition, the extent of the bidirectional blood shunt is known, adding yet another (unexplored) level of complexity to the control and regulation system.

There are stretch receptors in the aerial exchange organ of many air-breathing fish, amphibians, and reptiles, which can mediate profound ventilatory and cardiovascular responses in the absence of changes in gas partial pressures (Johansen 1970; Emilio and Shelton 1972; Johansen et al. 1977). Such a stretch receptor could supply afferent information on the volume of the bladder or lung, and hence monitor the buoyancy of the animal as well as the extent of the O_2 store remaining and its rate of removal during apnea. In the turtle, *Chrysemys scripta*, for example, pulmonary blood flow is proportional to lung volume independent of blood partial pressures (Figure 5.4), strongly suggesting that pulmonary stretch receptors are modulating lung perfusion to regulate and meter out the lung oxygen store.

Because the gas exchange ratio of the aerial exchange organ of bimodal breathers is often quite low (Figure 2.22, Chapter 2), the volume of the enclosed gas will fall progressively during apnea. A central integration of changes with respiration in the activity of tonically active stretch receptors could be important in producing the often complex changes in lung perfusion that can develop during diving (Figure 3.2, Chapter 3).

Figure 5.3. Relationship between the duration of breath holding (dive) and arterial P_{O_2} at the end of a period of pure N_2 breathing in the turtle *Chelys fimbriata*. As P_{aO_2} decreases, the duration of the dive also decreases (descending curves). Then at resumption of air breathing, the duration of the dive increases sharply (ascending curves) (from Lenfant et al., 1970).

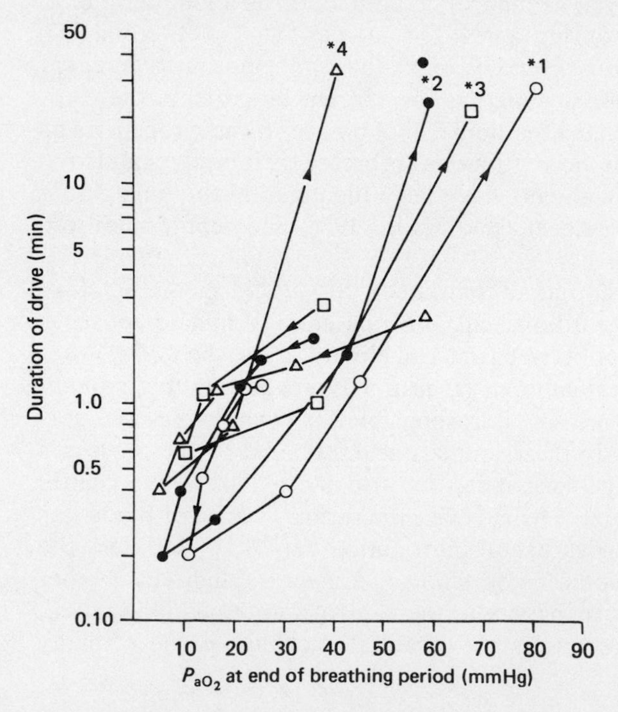

The breathing patterns of reptiles are variable, consisting of one or several breaths taken in sequence followed by an interbreath interval of variable duration. Feedback from lung stretch receptors is important in controlling tidal volume, which is held at an optimal value to minimize the cost of breathing (Milsom 1978). Reptiles increase lung ventilation by increasing the number of breaths taken, rather than altering tidal volume. Thus, in any given species, tidal volume is stereotyped and maintained at a given level by a series of feedback systems, and increased lung ventilation is achieved by initiating more breaths. How then is a breath initiated in an intermittently breathing vertebrate? In a great number of these animals hyperoxia prolongs, whereas hypoxia shortens, the interbreath interval, implying that it is the change in oxygen level of the "lung" gas or blood that governs the frequency of intermittent breathing (see Glass et al. 1978). In the intermittent breather *Chelys fimbriata,* oxygen levels in the blood or lung appear to determine the onset of breathing (Figure 5.3). There is no reason, however, to assume that all intermittent breathers should have similar control systems, or that oxygen is the sole stimulus, even in *Chelys,* especially in the light of the tremendous anatomical, physiological, and environmental diversity that exists in this group of animals.

Those oxygen receptors identified in vertebrates respond to hypoxia with an increase in firing rate. The inhibition of breathing in response to hyperoxia would indicate the removal of a tonic or phasic excitatory input to the respiratory neurons initiating a breath, whereas hypoxia would result in reflex excitation. The effects of oxygen on breathing are variable, however, even in a single animal, indicating that either the gain or the threshold level of the receptor system can be reset centrally or peripherally, though these and related problems in bimodal, intermittent breathers have received scant attention.

Figure 5.4. Changes in left pulmonary artery blood flow during stepwise withdrawals of gas from the lungs of *Chrysemys*. Replacement of the lung gas promptly returns heart rate and pulmonary blood flow to prewithdrawal values.

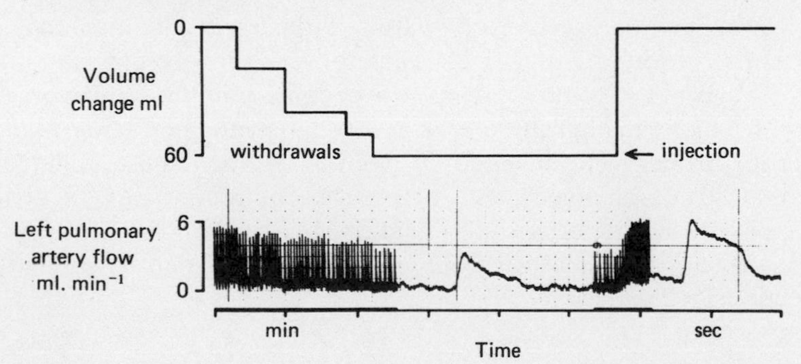

Continuous air breathers

Active terrestrial vertebrates probably evolved endothermy and the associated increase in metabolism as a result of selection pressures operating in temperate regions with low temperatures. Large oxygen stores, contained principally in the body fluids and constantly replenished by continuous lung ventilation, were selected for, but the rate of O_2 utilization from these stores was also rapid to support the burgeoning metabolic machinery required to maintain endothermy. Thus, the ratio of oxygen utilization to oxygen storage in continuous air breathers is low, and once again the careful balancing of gas supply and demand becomes critical. Although tremendous changes in nearly every aspect of respiration have occurred with the evolution of aspiration in the fully terrestrial vertebrates, these animals have in effect come full circle from the archetype water breathers, in the sense that they too must breathe continuously in order to maintain oxygen delivery to the tissues. The only functional difference is in the pattern of CO_2 excretion and acid–base regulation.

Compared to the bimodal, intermittent breathers the regulation of gas exchange is a simple process in the continuous air breathers. A single respiratory organ (the lung) is ventilated with a single respiratory medium (air), and the O_2 and CO_2 partial pressures of air are very stable and predictable. Afferent activity from highly sensitive peripheral arterial and central chemoreceptors is integrated in the brain to mediate ventilatory and cardiovascular responses to changes in blood gas levels which result from changes in supply or demand. Pulmonary afferent information also is an integral part of the regulation of ventilation. Different terrestrial vertebrates, however, have selected for different types of sensitivity from the archetype chemosensitive mechanoreceptor (Milsom and Jones 1976). Birds have evolved constant-volume lungs where mainly chemoreceptor activity is available to the central nervous system, whereas in tidally ventilated mammalian lungs proprioceptive information plays a fundamental role in the ongoing process of ventilation.

Most ventilatory and circulatory adjustments in mammals, under normal conditions, occur to regulate very closely plasma and tissue CO_2 and pH. Should O_2 transport become limiting, however, such as with a move to high altitude, then ventilation and perfusion adjustments occur primarily to enhance O_2 uptake, and acid–base balance in the short term is allowed to shift from the optimum. Thus oxygen would appear to be important in regulation, but in tetrapods it seems that a control system monitoring CO_2 and pH and modulating ventilation is of equal significance.

6 *The evolution of air breathing: a synthesis and summary*

We envisage that air-breathing vertebrates have evolved several times from water-breathing, bony fish belonging to the group Osteichthyes. These fish have within their body an enclosed chamber of air, which may function in buoyancy or as a source for oxygen. It is difficult to ascertain the evolutionary relationships between lungs and swim bladders, but they probably evolved independently, with lungs appearing before swim bladders. Because of their ventral location lungs have probably always functioned only in oxygen transfer, whereas dorsal swim bladders could function in either gas transfer or buoyancy control. These two functions are not mutually exclusive in swim bladders because the removal of oxygen without a simultaneous addition of an equal amount of CO_2 to the gas chamber will reduce its size and result in an increase in the overall density of the fish. In all probability this dorsal gas bladder initially served both as an oxygen source during aquatic hypoxia, and for buoyancy regulation, depending on the nature of the environment and the requirements of the animal. Subsequent evolution in fish resulted in the development of some forms specialized for buoyancy control and others in which modifications for more efficient gas transfer were selected. The latter would have involved the elaboration of a large, thin respiratory epithelium and some means of ventilation and perfusion of this epithelium.

Modern-day bimodal breathers may use a modified gill, the buccal or pharyngeal cavity, a region of the gut, or a lung or gas bladder for gas exchange. With the exception of a few specialized forms, which have surmounted ion regulation as well as gas exchange problems on land, most air-breathing fish are restricted to an aquatic existence; that is, they live in water and rise to the surface to tap the oxygen-rich atmosphere. In the terrestrial vertebrates, which arose from the few but successful air-breathing fish that penetrated onto land, the lung as a gas exchange organ was strongly selected for, and remains the major gas exchanger of tetrapods.

In all air-breathing organs, the size of the respiratory surface and the extent of perfusion are correlated with the dependence of the animal on air breathing. In view of (1) a usually very thin blood–gas barrier (0.1 – 15.0 μm) in a wide range of aerial exchange organs, (2) the rapid initial depletion of O_2 from gas exchangers during apnea, and (3) physiological evidence showing near-equilibration between gas and the blood in aerial exchangers (West 1977; Burggren and Shelton 1979), we strongly suspect that gas transfer across all vertebrate air-breathing organs is perfusion-, rather than diffusion-, limited. Consequently, elaborate mechanisms for controlling perfusion of the aerial exchanger have evolved to regulate gas transfer, particularly when this organ is located in parallel with the systemic circulation.

Vertebrate air-breathing organs are ventilated by either a buccal pump or aspiration. The buccal pump, used for ventilating the gills of aquatic fish with water, has been modified in amphibians and many air-breathing fish to force air into the air-breathing organ. These modifications may be functional, as, for instance, in the timing and contribution of the contraction of different muscle groups to buccal contraction. Structural modifications may also occur (through evolution or ontogeny) in which, for example, opercular movements are uncoupled from buccal movements to maintain opercular closure as the buccal cavity contracts and air is forced into the air-breathing organ.

The use of a buccal pump in air breathing places restrictions on jaw structure and, therefore, feeding behavior. This is not the case for animals that have evolved an aspiratory mode of lung ventilation. In those animals the mouth and/or nares are still passages for air flow, but the jaws and buccal cavity in general play little or no active role in ventilation. The jaws can then be modified for feeding requirements alone.

In all probability both buccal pumping and aspiration to ventilate the air-breathing organ evolved several times in animals living in water. The buccal pump is used to ventilate the gas bladder of many air-breathing fish, whereas the entirely aquatic but obligatory air-breathing teleost *Arapaima* ventilates its well-developed lung by aspiration. Thus, we conclude that both buccal pumping and aspiration breathing evolved in animals living in water but breathing air.

The simplest modification is that associated with the buccal pump, and we suspect that animals using buccal ventilation of their lung evolved onto land first. These animals, with relatively light but large jaws, were probably herbivorous, with limited predatory ability. They were, however, subsequently followed onto land by the carnivorous, heavier-jawed aspiratory-breathing vertebrates. Both of these groups probably retained gills, and the lung was used mainly for oxygen uptake, whereas the gills were used for CO_2, acid–base, and ion regulation; that is, they retained a pattern of pH regulation characteristic of aquatic animals. Thus, these animals could only spend short periods of time actively moving about on land.

The reduction of gills and the relocation of the CO_2, [H^+], and ion regulatory functions of the gills to a moist skin, plus a variety of water-conserving or bulk-transporting devices (e.g., toads carry an enormous volume of water in their bladder when they leave the water), allowed animals greater penetration onto land. These air-breathing vertebrates, however, like the modern amphibians, were ultimately tied to water for acid–base and ion regulation. Therefore, there were strong selective forces operating in these primitive terrestrial vertebrates for the evolution of alternate mechanisms for CO_2 and acid–base regulation while on land. One such mechanism was to excrete CO_2 via the lung. This required modifications in the properties of erythrocytes (Chapter 2), which must have occurred in the ancestral group of the modern reptiles, birds, and mammals. In this group plasma bicarbonate could be catalyzed to CO_2 during a passage through the lung, and the bicarbonate dehydration reaction was no longer the limiting factor in lung CO_2 excretion. Carbon dioxide excretion was limited, and therefore could be controlled, by lung ventilation.

Precise control of lung ventilation is afforded by aspiratory ventilation in which both rate and tidal volume can be regulated. A moist skin is no longer required, and the body surface can be dry, impermeable to water, and tough. The result is a reduction in water loss, which then becomes related to the degree of lung ventilation. These animals and their descendants, the modern reptiles, birds, and mammals, can remain active in fairly arid environments because they have a much-reduced dependence on a continuous supply of water compared with air-breathing fish and amphibians.

Thus the evolution of reptiles, birds, and mammals from aquatic ancestors probably first involved the evolution of a lung, ventilated by aspiration and utilized for oxygen uptake. This bimodally breathing progenitor lived in water and retained gills for CO_2 excretion and H^+ regulation. Further evolution may have relocated these gill functions in the skin and resulted in the loss of gills. Regardless of whether the gills were lost, these animals were restricted to only short periods of movement on land. Subsequently, a mechanism for CO_2 elimination via the lung evolved, and the lung became the only respiratory organ.

These changes from a unimodal water-breathing, to a bimodal and then unimodal air-breathing animal have involved marked changes in the circulation. Initially in series in water breathers, the blood supply to the gas exchanger(s) evolved to become arranged in parallel with the systemic circuit in most bimodal breathers, and in unimodal air breathers is once again arranged in series with the circulation to the body tissues. Most extant, noncrocodilian reptiles and amphibians have an anatomically and functionally undivided ventricle that pumps blood to both the pulmonary (pulmocutaneous) and systemic circulations. The volume of blood pumped into one circuit need not be equal to that entering the other as it must be in adult birds and mammals, and the absolute flow to either circuit, as well as the admixture of arterial and venous blood, can be varied, depending on the relative impedance of the two circulations. Most amphibians and reptiles are intermittent breathers, and have retained an undivided circulation as a means of redistributing blood between pulmonary and systemic circuits between breaths. Many of these animals are aquatic, and have large lung volumes and therefore large oxygen stores, sufficient to supply their oxygen requirements for several minutes, or even hours, while they remain submerged. They regulate lung perfusion somewhat independently of systemic flow to control the rate of oxygen removal from the lung store. This is only possible with an undivided circulation.

Many reptiles have evolved into fully terrestrial forms. Some snakes and tortoises still breathe intermittently while on land, probably as a consequence of their low metabolic rates (Chapter 3), and have cardiovascular systems identical to their aquatic relatives. Other terrestrial reptiles, such as some snakes and the lizards, tend toward a continuous breathing pattern, yet have retained an undivided circulation (Figure 6.1). The selective advantage of retaining a functionally undivided ventricle in this case is not clear. Lizards can apparently alter the balance between pulmonary and systemic blood flow for temperature regulation. These animals control heat transfer by perfusing the skin

Figure 6.1. The functional evolution of aspiratory breathing patterns in terrestrial vertebrates.

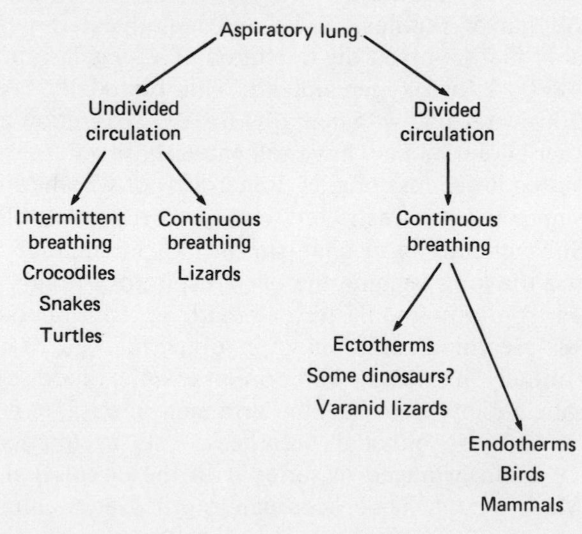

with varying amounts of deoxygenated systemic venous blood that is recirculated directly to the skin without first passing through the pulmonary circulation. In this case the undivided circulation serves heat regulation at the expense of optimal gas transfer.

The varanid lizard is unusual among extant reptiles in that it has a functionally divided circulation during a portion of the cardiac cycle, capable of perfusing the systemic and pulmonary circuits at two quite different pressures (Millard and Johansen 1974), as in adult birds and mammals. It has been suggested that some large dinosaurs may also have possessed a divided circulation (Marx 1978). Certainly, high pressures must have been developed in the ventricle in order to raise blood to the head of *Brontosaurus*, for example, which must have been many meters above the heart when held erect. It seems unlikely that these high systemic arterial pressures were transmitted to the fragile lung capillaries; so either the ventricle was divided, or there was a large resistance in the pulmonary artery.

Debate surrounds the question of whether the dinosaurs were ectotherms, like modern reptiles, or endotherms, like birds and mammals, the groups that blossomed following the demise of the dinosaurs. Modern reptiles (e.g., lizards) are ectothermic, and one wonders what selective advantage these animals have over mammals that has enabled them to survive. Endotherms have much higher rates of energy utilization, and therefore a higher food-intake requirement. Lizards can move quickly, but are more dependent on the ambient temperature. At the same time they have much lower energy requirements and can reduce energy consumption to very low levels during the night or in winter. Thus they can survive on a very low food ration. This gives them a selective advantage over the energy-consuming birds and mammals. Certainly the majority of reptiles exist in warmer climates where the advantages of ectothermy over endothermy are less apparent.

Thus the evolution of air-breathing vertebrates has been associated with marked changes in both the structure and the location of the gas exchange organ, its mode of ventilation, and the means of blood perfusion. These changes have occurred in a series of steps involving the change from unimodal water breathing to bimodal breathing to unimodal air breathing. The intermediate bimodal stage occurred because of the temporal separation of the evolution of the use of an air-breathing organ for oxygen uptake and the subsequent changes in the pattern of CO_2 excretion and acid–base regulation.

References

Baglioni, S. 1907. Der Atmungsmechismus der Fishe. Ein Beitrag zur vergleichenden Physiologie des Atemrhythms. *Z. Allg. Physiol.* 7: 177–282.

Ballintijn, C. M. and G. M. Hughes. 1965. The muscular basis of the respiratory pumps in the trout. *J. Exp. Biol.* 43: 349–62.

Bamford, O. S. 1974. Oxygen reception in the rainbow trout (*Salmo gairdneri*). *Comp. Biochem. Physiol.* 48A: 69–76.

Bartlett, G. R. 1978a. Water-soluble phosphates of fish red cells. *Can. J. Zool.* 56: 870–7.

– 1978b. Phosphates in red cells of two South American osteoglossids: *Arapaima gigas* and *Osteoglossum bicirrhosum*. *Can. J. Zool.* 56: 878–81.

– 1978c. Phosphates in red cells of two lungfish: the South American, *Lepidosiren paradoxa*, and the African, *Protopterus aethiopicus*. *Can. J. Zool.* 56: 882–6.

Bastert, C. 1929. Respiration and circulation of the blood of *Rana esculenta* and *Rana fusca* in connection with their diving habits. *Tijdschr. Ned. Dierk. Ver.* 1: 98–104.

Belkin, D. A. 1964: Variations in heart rate during voluntary diving in the turtle *Pseudemys concinna*. *Copeia* 2: 321–30.

Berg, T. and J. B. Steen. 1965. Physiological mechanisms for aerial respiration in the eel. *Comp. Biochem. Physiol.* 15: 469–84.

– 1966. Regulation of ventilation in eels exposed to air. *Comp. Biochem. Physiol.* 18: 511–16.

– 1968. The mechanism of oxygen concentration in the swimbladder of the eel. *J. Physiol.* 195: 631–8.

Berger, P. J. 1972. The vagal and sympathetic innervation of the isolated pulmonary artery of a lizard and a tortoise. *Comp. Gen. Pharmacol.* 3: 113–24.

Bettex-Galland, M. and G. M. Hughes. 1973. Contractile filamentous material in the pillar cells of fish gills. *J. Cell Sci.* 13: 359–70.

Bolis, L. and P. Luly. 1972. Monosaccharide permeability in brown trout, *Salmo trutta* L. erythrocytes. In *Role of Membranes in Secretory Processes*, ed. L. Bolis, R. D. Keynes, and W. Wilbrandt, pp. 215–21. London and Amsterdam: North Holland Co.

Booth, J. H. 1978. The distribution of blood flow in the gills of fish: Application of a new technique to rainbow trout (*Salmo gairdneri*). *J. Exp. Biol.* 73: 119–29.

Bornancin, M., G. De Renzis, and J. Maetz. 1977. Branchial Cl^- transport anion stimulated ATPase and acid–base balance in *Anguilla anguilla* adapted to freshwater: Effects of hyperoxia. *J. Comp. Physiol. B* 117: 313–22.

Boutilier, R. G., D. J. Randall, G. Shelton, and D. P. Toews. 1979a. Acid–base relationships in the blood of the toad, *Bufo marinus*. I. The effects of environmental CO_2. *J. Exp. Biol.* 82: 331–44.

– 1979b. Acid–base relationships in the blood of the toad, *Bufo marinus*. II. The effects of dehydration. *J. Exp. Biol.* 82: 345–55.

Burg, M. and N. Green. 1977. Bicarbonate transport by isolated perfused rabbit proximal convoluted tubules. *Am. J. Physiol.* 233: F307–14.

Burggren, W. W. 1975. A quantitative analysis of ventilation tachycardia and its control in two chelonians *Pseudemys scripta* and *Testudo graeca*. *J. Exp. Biol.* 63: 367–80.

– 1977a. The pulmonary circulation of the chelonian reptile: Morphology, haemodynamics and pharmacology. *J. Comp. Physiol.* 116: 303–23.

– 1977b. Circulation during intermittent lung ventilation in the garter snake *Thamnophis*. *Can. J. Zool.* 55: 1720–5.

– 1978. Influence of intermittent breathing on ventricular depolarization patterns in chelonian reptiles. *J. Physiol.* 278: 349–64.

– 1979. Bimodal gas exchange during variation in environmental oxygen and carbon dioxide in the air breathing fish *Trichogaster trichopterus*. *J. Exp. Biol.*, 82: 197–213.

Burggren, W. W. and S. Haswell. 1979. Aerial CO_2 excretion in the obligate air breathing fish, *Trichogaster trichopterus:* A role for carbonic anhydrase. *J. Exp. Biol.*, 82: 215–260.

Burggren, W. W. and D. J. Randall. 1978. Oxygen uptake and transport during hypoxic exposure in the sturgeon *Acipenser transmontanus*. *Respir. Physiol.* 34: 279–91.

Burggren, W. W. and G. Shelton. 1979. Gas exchange and transport during intermittent breathing in chelonian reptiles. *J. Exp. Biol.*, 82: 75–92.

Burggren, W. W., M. L. Glass, and K. Johansen. 1977. Pulmonary ventilation/perfusion relationships in terrestrial and aquatic chelonian reptiles. *Can. J. Zool.* 55: 2024–34.

Cameron, J. N. 1976. Branchial ion uptake in Arctic grayling: Resting values and effects of acid–base disturbance. *J. Exp. Biol.* 64: 711–25.

– 1978a. Regulation of blood pH in teleost fish. *Respir. Physiol.* 33: 129–44.

– 1978b. Chloride shift in fish blood (1). *J. Exp. Zool.* 206: 289–95.

Cameron, J. N. and D. J. Randall. 1972. The effect of increased ambient CO_2 on arterial CO_2 tension, CO_2 content and pH in rainbow trout. *J. Exp. Biol.* 57: 673–80.

Cameron, J. N. and C. M. Wood. 1978. Renal function and acid–base regulation in two Amazonian erythrinid fishes: *Hoplias malabaricus*, a water breather, and *Hoplerythrinus unitaeniatus*, a facultative air breather. *Can, J. Zool.* 56: 917–30.

Campbell, G. 1971. Autonomic innervation of the pulmonary vascular bed in a toad (*Bufo marinus*). *Comp. Gen. Pharmacol.* 2: 287–94.

Carter, G. S. and L. C. Beadle. 1931. The fauna of the swamps of the

Paraguayan Chaco in relation to its environment. II. Respiratory adaptations in the fishes. *J. Linn. Soc. Lond.* 37: 327–66.

Clark, B. D., C. Gans, and H. I. Rosenburg. 1978. Air flow in snakes. *Respir. Physiol.* 32: 207–12.

Cloud, P. 1974. Evolution of ecosystems. *Am. Sci.* 62: 54–66.

Das, B. K. 1927. III. The bionomics of certain air breathing fishes of India, together with an account of the development of their air-breathing organs. *Phil. Trans. Roy. Soc. Lond.* 216: 183–219.

– 1934. The habits and structure of *Pseudapocrytes lanceolatus*, a fish in the first stages of structural adaptation to aerial respiration. *Proc. Roy. Soc. B.* 115: 422–30.

Daxboeck, C. and G. F. Holeton. 1978. Oxygen receptors in the rainbow trout, *Salmo gairdneri*. *Can. J. Zool.* 56: 1254–9.

Dehadrai, P. V. and S. D. Tripathi. 1976. Environment and ecology of freshwater air-breathing teleosts. In *Respiration of Amphibious Vertebrates,* ed. G. H. Hughes, pp. 39–72. London: Academic Press.

Dejours, P. 1973. Problems of control of breathing in fishes. In *Comparative Physiology*, ed. L. Bolis, K. Schmidt-Nielsen, and S. H. P. Maddrell, pp. 117–33. Amsterdam: North Holland Co.

DeLaney, R. G. and A. P. Fishman. 1977. Analysis of lung ventilation in the aestivating lungfish *Protopterus aethiopicus*. *Am. J. Physiol.* 233: R181–7.

DeLaney, R. G., S. Lahiri, and A. P. Fishman. 1974. Aestivation of the African lungfish *Protopterus aethiopicus*: Cardiovascular and respiratory functions. *J. Exp. Biol.* 61: 111–8.

DeLaney, R. G., S. Lahiri, R. Hamilton, and A. P. Fishman. 1977. Acid–base status and plasma composition in the aestivating lungfish (*Protopterus*). *Am. J. Physiol.* 232: R10–17.

Dence, W. A. 1933. Notes on a large bowfin (*Amia calva*) living in a mud puddle. Copiea, Ichthyological Notes 1: 35.

De Renzis, G. and J. Maetz. 1973. Studies on the mechanism of chloride absorption by the goldfish gill: Relation with acid–base regulation. *J. Exp. Biol.* 59: 339–58.

Effros, R. M., R. S. Y. Chang, and P. Silverman. 1978. Acceleration of plasma bicarbonate conversion to carbon dioxide by pulmonary carbonic anhydrase. *Science* 199: 427–9.

Elancher, B. 1974. Contrôle de le respiration chez les poissons téléostéens: Réactions respiratoires à des changements rectangulaires de l' oxygénation du milieu. *C. R. Acad. Sci. Paris* 280: 307–10.

Elancher, B. and P. Dejours. 1974. Contrôle de la respiration chez les poissons téléostéens: Existence de chémorécepteurs physiologiquement analogues aux chémorécepteurs des Vertébrés supérieurs. *C. R. Acad. Sci. Paris* 280: 451–3.

Emilio, M. G. and G. Shelton. 1972. Factors affecting blood flow to the lungs in the amphibian, *Xenopus laevis*. *J. Exp. Biol.* 56: 67–77.

– 1974. Gas exchange and its effect on blood gas concentrations in the amphibian, *Xenopus laevis*. *J. Exp. Biol.* 60: 567–79.

Emilio, M. G., M. M. Machado, and H. P. Menano. 1970. The production of a

hydrogen ion gradient across the isolated frog skin. Quantitative aspects and the effect of acetazolamide. *Biochim. Biophys. Acta* 203: 394–409.
Farber, J. and H. Rahn. 1970. Gas exchange between air and water and the ventilation pattern in the electric eel. *Respir. Physiol.* 9: 151–61.
Farmer, M. 1979. The transition from water to air breathing: Effects of CO_2 on hemoglobin function. *Comp. Biochem. Physiol.* 62A: 109–14.
Farrell, A. P. 1978. Cardiovascular events associated with air breathing in two teleosts, *Hoplerythrinus unitaeniatus* and *Arapaima gigas*. *Can. J. Zool.* 56: 953–8.
– 1979. Gill blood flow in teleosts. Ph.D. thesis, University of British Columbia.
Farrell, A. P. and D. J. Randall. 1978. Air-breathing mechanics in two Amazonian teleosts, *Arapaima gigas* and *Hoplerythrinus unitaeniatus*. *Can. J. Zool.* 56: 939–45.
Fisher, T. R., R. F. Coburn, and R. E. Forster. 1969. Carbon monoxide diffusing capacity in the bullhead catfish. *J. Appl. Physiol.* 26: 161–9.
Forster, R. E. and E. D. Crandall. 1975. Time course of exchanges between red cells and extracellular fluid during CO_2 uptake. *J. Appl. Physiol.* 38: 710–18.
Forster, R. E. and J. B. Steen. 1969. The rate of the "Root shift" in eel red cells and eel hemoglobin solutions. *J. Physiol.* 204: 259–82.
Foxon, G. E. H. 1964. Cardiac physiology of a urodele amphibian. *Comp. Biochem. Physiol.* 13: 47–53.
Gans, C. 1970. Strategy and sequence in the evolution of the external gas exchangers of ectothermal vertebrates. *Forma Functio* 3: 61–104.
– 1976. Ventilatory mechanisms and problems in some amphibious aspiration breathers (*Chelydra, Caiman* – Reptilia). In *Respiration of Amphibious Vertebrates*, ed. G. M. Hughes, pp. 357–74. London: Academic Press.
Gans, C. and B. Clark. 1976. Studies on ventilation of *Caiman crocodilis* (Crocodilia: Reptilia). *Respir. Physiol.* 26: 285–301.
– 1978. Air flow in reptilian ventilation. *Comp. Biochem. Physiol.* 60A: 453–7.
Gans, C. and G. M. Hughes. 1967. The mechanism of lung ventilation in the tortoise *Testudo graeca* Linné. *J. Exp. Biol.* 47: 1–20.
Gaunt, A. S. and C. Gans. 1969. Mechanics of respiration in the snapping turtle *Chelydra serpentina* (Linné). *J. Morphol.* 128: 195–227.
Gee, J. H. and J. P. Graham. 1978. Respiratory and hydrostatic functions of the intestine of the catfishes *Hoplosternum thoracatum* and *Brochis splendens* (Callichthyidae). *J. Exp. Biol.* 74: 1–16.
Gerald, J. W. and J. J. Cech. 1970. Respiratory responses of juvenile catfish (*Ictalurus punctatus*) to hypoxic conditions. *Physiol. Zool.* 43: 47–54.
Giebisch, G., L. Berger, and R. F. Pitts. 1955. The extrarenal response to acute acid–base disturbances of respiratory origin. *J. Clin. Invest.* 34: 231–45.
Glass, M., W. Burggren, and K. Johansen. 1978. Ventilation and its control in an aquatic and a terrestrial chelonian reptile. *J. Exp. Biol.* 72: 165–80.
Graham, J. B. 1973. Terrestrial life of the amphibious fish *Mnierpes macrocephalus*. *Mar. Biol.* 23: 83–91.
Graham, J. B., D. L. Kramer, and E. Pineda. 1977. Respiration of the air breathing fish *Piabucina festae*. *J. Comp. Physiol. B* 122: 295–310.
Gregory, R. B. 1977. Synthesis and total excretion of waste nitrogen by fish of

the *Perioptha lmus* (mudskipper) and Scartelaos families. *Comp. Biochem. Physiol.* 57A: 33–6.
Gros, G. and W. Moll. 1974. Facilitated diffusion of CO_2 across albumin solutions. *J. Gen. Physiol.* 64: 356–71.
Gulliver, G. 1875. Observations on the sizes and shapes of the red corpuscles of the blood of vertebrates, with drawings of them to uniform scale, and extended and revised tables of measurements. *Proc. Zool. Soc. Lond.* 474–95.
Hansen, V. K. and K. G. Wingstrand. 1960. Further studies on the non-nucleated erythrocytes of *Maurolieus mulleri,* and comparisons with the blood cells of related fish. Dana report No. 54. Copenhagen: A. F. Host and Sons. 15 pp.
Hanson, D. 1967. Cardiovascular dynamics and aspects of gas exchange in Chondricthyes. Ph.D. thesis, University of Washington.
Haswell, M. S. 1978. CO_2 excretion and acid–base regulation in the rainbow trout, *Salmo gairdneri*. Ph.D. thesis, University of British Columbia.
Haswell, M. S. and D. J. Randall. 1976. Carbonic anhydrase inhibitor in trout plasma. *Respir. Physiol.* 28: 17–27.
– 1978. The pattern of carbon dioxide excretion in the rainbow trout *Salmo gairdneri. J. Exp. Biol.* 72: 17–24.
Haswell, M. S., R. Zeidler, and H. D. Kim. 1978. Chloride transport in red cells of a teleost, *Tilapia mossambica. Comp. Biochem. Physiol.* 61A: 217–20.
Haswell, M. S., S. F. Perry, and D. J. Randall. 1980. Fish gill carbonic anhydrase: Acid–base regulation or salt transport? *Am. J. Physiol.,* 238 (Regulatory Integrative Comp. Physiol. 7): R240–R245.
Hill, E. P., G. G. Power, and R. D. Gilbert, 1977. Rate of pH changes in blood plasma in vitro and in vivo. *J. Appl. Physiol.* 42: 928–34.
Hills, B. A. 1972. Diffusion and convection in lungs and gills. *Respir. Physiol.* 14: 105–14.
Holeton, G. F. and D. R. Jones. 1975. Water flow dynamics in the respiratory tract of the carp (*Cyprinus carpio* L.). *J. Exp. Biol.* 63: 537–49.
Holeton, G. F. and D. J. Randall. 1967. The effect of hypoxia upon the partial pressure of gases in the blood and water afferent and efferent to the gills of rainbow trout. *J. Exp. Biol.* 46: 317–27.
Holland, H. D. 1975. The evolution of seawater. In *The Early History of the Earth,* ed. B. F. Windley, pp. 559–67. London and New York: John Wiley and Sons.
Holser, P. 1977. Catastrophic chemical events in the history of the oceans. *Nature* 267: 403–8.
Hughes. G. M. 1960. A comparative study of gill ventilation in marine teleosts. *J. Exp. Biol.* 37: 28–45.
– 1964. How a fish extracts oxygen from water. *New Sci.* 11: 346–8.
Hughes, G. M. and G. Shelton. 1957. Pressure changes during the respiratory movements of teleostian fishes. *Nature* 179: 255–7.
Isaacks, R. E., H. D. Kim, and D. R. Harkness. 1978. Relationship between phosphorylated metabolic intermediates and whole blood oxygen affinity in some air-breathing and water-breathing teleosts. *Can. J. Zool.* 56: 887–90.
Janes, R. 1979. Hyperoxia in the toad *Bufo marinus*. Honours undergraduate thesis, Acadia University.
Janssen R. G. and D. J. Randall. 1975. The effects of changes in pH and P_{CO_2} in

blood and water on breathing in rainbow trout, *Salmo gairdneri*. *Respir. Physiol.* 25: 235–45.
Johansen, K. 1966. Chemoreception in respiratory control of lungfish, *Neoceratodus*. *Fed. Proc.* 25: 389.
– 1970. Air breathing in fishes. In *Fish Physiology*, ed. W. S. Hoar and D. J. Randall, vol. 4, pp. 361–411. New York: Academic Press.
Johansen, K. and D. Hanson. 1968. Functional anatomy of the hearts of lungfishes and amphibians. *Am. Zool.* 8: 191–210.
Johansen, K. and R. Hol. 1968. A radiological study of the central circulation in the lungfish, *Protopterus aethiopicus*. *J. Morphol.* 126: 333–48.
Johansen, K. and C. Lenfant. 1967. Respiratory function in the South American lungfish, *Lepidosiren paradoxa* (Fitz). *J. Exp. Biol.* 46: 205–18.
– 1968. Respiration in the African lungfish *Protopterus aethiopicus*. II. Control of breathing. *J. Exp. Biol.* 49: 453–68.
Johansen, K., C. Lenfant, and D. Hanson. 1968a. Cardiovascular dynamics in the lungfishes. *Z. Vgl. Physiol.* 59: 157–86.
Johansen, K., C. Lenfant, K. Schmidt-Nielson, and J. A. Petersen. 1968b. Gas exchange and control of breathing in the electric eel, *Electrophorus electricus*. *Z. Vgl. Physiol.* 61: 137–63.
Johansen, K., D. Hanson, and C. Lenfant. 1970. Respiration in the primitive air breather, *Amia calva*. *Respir. Physiol.* 9: 162–74.
Johansen, K., W. Burggren, and M. Glass. 1977. Pulmonary stretch receptors influence pulmonary blood flow and heart rate in the turtle, *Pseudemys scripta*. *Comp. Biochem. Physiol.* 58A: 185–91.
Johansen, K., C. P. Mangum, and G. Lykkeboe. 1978. Respiratory properties of the blood of Amazon fishes. *Can. J. Zool.* 56: 898–906.
Jones, D. R. 1967. Oxygen consumption and heart rate of several species of anuran Amphibia during submergence. *Comp. Biochem. Physiol.* 20: 691–707.
Jones, D. R. and D. J. Randall. 1978. The respiratory and circulatory systems during exercise. In *Fish Physiology*, ed. W. S. Hoar and D. J. Randall, vol. 7, pp. 425–501. New York: Academic Press.
Jones, D. R., D. J. Randall, and G. M. Jarman. 1970. A graphic analysis of oxygen transfer in fish. *Respir. Physiol.* 10: 285–98.
Kerstetter, T. H. and L. B. Kirschner. 1972. Active chloride transport by the gills of rainbow trout (*Salmo gairdneri*). *J. Exp. Biol.* 56: 263–72.
Kim, H. D. and R. E. Isaacks. 1978. The membrane permeability of nonelectrolytes and carbohydrate metabolism of Amazon fish red cells. *Can. J. Zool.* 56: 863–9.
Kirschner, L. B., L. Greenwald, and T. H. Kerstetter. 1973. Effect of amiloride on sodium transport across the body surfaces of freshwater animals. *Am. J. Physiol.* 224: 832–7.
Kramer, D. L. 1978. Ventilation of the respiratory gas bladder in *Hoplerythrinus unitaeniatus* (Pisces, Characoidei, Erythrinidae). *Can. J. Zool.* 56: 931–8.
Laurent, P., R. G. DeLaney, and A. P. Fishman. 1978. The vasculature of the gills in the aquatic and estivating lungfish (*Protopterus aethiopicus*). *J. Mophol.* 156: 173–208.
Lefant, C. and K. Johansen. 1968. Respiration in the African lungfish, *Protop-*

terus aethiopicus. I. Respiratory properties of blood and normal patterns of gas exchange. *J. Exp. Biol.* 49: 437–52.
Lefant, C., K. Johansen, J. A. Petersen, and K. Schmidt-Nielsen. 1970. Respiration in the fresh water turtle, *Chelys fimbriata*. *Respir. Physiol.* 8: 261–75.
McDonald, H. S. 1959. Respiratory functions of the ophiodon air sac. *Herpteologica* 15: 193–8.
Macintyre, D. H. and D. P. Toews. 1976. The mechanics of lung ventilation and the effects of hypercapnia on respiration in *Bufo marinus*. *Can. J. Zool.* 54: 1364–74.
McLean, J. R. and G. Burnstock. 1967. Innervation of the lungs of the toad (*Bufo marinus*). II. Flourescent histochemistry of catecholamines. *Comp. Biochem. Physiol.* 22: 767–73.
McMahon, B. R. 1969. A functional analysis of the aquatic and aerial respiratory movements of an African lungfish, *Protopterus aethiopicus*, with reference to the evolution of the lung-ventilation mechanism in vertebrates. *J. Exp. Biol.* 51: 407–30.
Maetz, J. 1973. Na^+/NA_4^+, Na^+/H^+ exchanges and NH_3 movements across the gill of *Carassius auratus*. *J. Exp. Biol.* 58: 255–75.
Maetz, J. and F. Garcia-Romeu. 1964. The mechanism of sodium and chloride uptake by the gills of a fresh-water fish, *Carassius auratus*. II. Evidence for NH_4^+/Na^+ and HCO_3^-/Cl^- exchanges. *J. Gen. Physiol.* 47: 1209–27.
Marx, J. L. 1978. Warm-blooded dinosaurs: Evidence pro and con. *Science* 199: 1424–6.
Millard, R. W. and K. Johansen. 1974. Ventricular outflow dynamics in the lizard, *Varanus niloticus:* Responses to hypoxia, hypercarbia and diving. *J. Exp. Biol.* 60: 871–80.
Milsom, W. K. 1978. Pulmonary receptors and their role in the control of breathing in turtles. Ph.D. thesis, University of British Columbia.
Milsom, W. K. and D. R. Jones. 1976. Reptilian pulmonary receptors: Mechano- or chemosensitive? *Nature* 261: 327–8.
– 1979. The role of pulmonary afferent information and hypercapnia in the control of the breathing pattern in chelonia. *Respir. Physiol.* 37: 101–7.
Moy-Thomas, J. A. 1971. *Palaeozoic Fishes,* 2nd ed., revised by R. S. Miles. Philadelphia: W. B. Saunders Co. 259 pp.
Obaid, A. L., A. McElroy Critz, and E. D. Crandell 1980. Kinetics of bicarbonate/chloride exchange in dogfish erythrocytes. *Am. J. Physiol.* In press.
Oduleye, S. O. 1977. Unidirectional water and sodium fluxes and respiratory metabolism in the African lungfish, *Protopterus annectens*. *J. Comp. Physiol.* 119: 127–39.
Olsen, C. R., R. Elsner, F. C. Hale, and D. W. Kenney. 1969. "Blow" of the pilot whale. *Science* 163: 953–5.
Packard, G. C. 1974. The evolution of air-breathing in Paleozoic gnathostome fishes. *Evolution* 28: 320–5.
Palmer, A. R. 1974. Search for the Cambrian world. *Am. Sci.* 62: 216–24.
Payan, P. and J. Maetz. 1973. Branchial sodium transport mechanisms in *Scyliorhinus canicula:* Evidence for Na^+/NH_4^+ and Na^+/H^+ exchanges and for a role of carbonic anhydrase. *J. Exp. Biol.* 58: 487–502.
Peters, H. M. 1978. On the mechanism of air ventilation in anabantoids (Pisces: Teleostei). *Zoomorphologie* 89: 93–123.

Phleger, C. F. and B. S. Saunders. 1978. Swim-bladder surfactants of Amazon air-breathing fishes. *Can. J. Zool.* 56: 946–52.

Potts, W. T. W. and P. P. Rudy. 1972. Aspects of osmotic and ionic regulation in the sturgeon. *J. Exp. Biol.* 56: 703–15.

Powers, D. A., H. J. Fyhn, U. E. H. Fyhn, J. P. Martin, R. L. Garlick, and S. C. Wood. 1979. A comparative study of the oxygen equilibria of blood from 40 genera of Amazonian fishes. *Comp. Biochem. Physiol.* 62A: 67–86.

Rahn, H. and W. F. Garey. 1973. Arterial CO_2, O_2, pH and HCO_3^- values of ectotherms living in the Amazon. *Am. J. Physiol.* 225: 735–38.

Rahn, H., K. B. Rahn, B. J. Howell, C. Gans, and S. M. Tenny. 1971. Air breathing of the garfish. *Lepisosteus osseus. Respir. Physiol.* 11: 443–66.

Randall, D. J. 1970a. Gas exchange in fish. In *Fish Physiology*. ed. W. S. Hoar and D. J. Randall, vol. 4, pp. 252–92. New York: Academic Press.

– 1970b. The circulatory system. In *Fish Physiology*. Ed. W. S. Hoar and D. J. Randall, vol. 4, pp. 133–72. New York: Academic Press.

Randall, D. J. and J. N. Cameron. 1973. Respiratory control of arterial pH as temperature changes in rainbow trout, *Salmo gairdneri*. *Am. J. Physiol.* 225: 997–1002.

Randall, D. J. and D. R. Jones. 1973. The effect of deafferentation of the pseudobranch on the respiratory response to hypoxia and hyperoxia in the trout (*Salmo gairdneri*). *Respir. Physiol.* 17: 291–301.

Randall, D. J. and L. S. Smith. 1967. The effect of environmental factors on circulation and respiration in teleost fish. *Hydrobiologia* 29: 113–24.

Randall, D. J., D. Baumgarten, and M. Malyusz. 1972. The relationship between gas and ion transfer across the gills of fishes. *Comp. Biochem. Physiol.* 41A: 629–37.

Randall, D. J., N. Heisler, and F. Drees. 1976. Ventilatory response to hypercapnia in the larger spotted dogfish *Scyliorhinus stellaris*. *Am. J. Physiol.* 230: 590–4.

Randall, D. J., A. P. Farrell, and M. S. Haswell. 1978a. Carbon dioxide excretion in the jeju *Hoplerythrinus unitaeniatus,* a facultative air-breathing teleost. *Can. J. Zool.* 56: 970–3.

Randall, D. J., A. P. Farrell, and M. S. Haswell. 1978b. Carbon dioxide excretion in the pirarucu (*Arapaima gigas*), an obligate air-breathing fish. *Can J. Zool.* 56: 977–82.

Reeves, R. B. and H. Rahn. 1979. Patterns in vertebrate acid–base regulation. In *Evolution of Respiratory Processes,* ed. S. C. Wood and C. Lenfant, pp. 225–52. New York: Marcel Dekker, Inc.

Reeves, R. B., B. J. Howell, and H. Rahn. 1977. *Protons, Proteins, Temperature*. Publications in acid–base physiology, Dept. of Physiology, State University of New York at Buffalo. 301 pp.

Riggs, A. 1979. Studies of the hemoglobins of Amazonian fishes: An overview. *Comp. Biochem. Physiol.* 62A: 257–71.

Rogers, D. C., and C. J. Haller. 1978. Innervation and cytohistochemistry of the neuroepithelial bodies in the ciliated epithelium of the toad lung (*Bufo marinus*). *Cell Tissue Res.* 195: 395–410.

Romer, A. S. 1970. *The Vertebrate Body*, 4th ed. Philadelphia: W. B. Saunders Co. 644 pp.

– 1972. Skin breathing – primary or secondary? *Respir. Physiol.* 14: 183–92.

Rosenburg, H. I. 1973. Functional anatomy of pulmonary ventilation in the garter snake, *Thamnophis elegans*. *J. Morphol.* 140: 171–84.
Rossi-Bernardi, L. and F. J. W. Roughton. 1970. The role of oxygen linked carbamate in the transport of CO_2 by human erythrocytes under physiological conditions. *J. Physiol. Lond.* 209: 25P–26P.
Sachs, G. 1977. Ion pumps in the renal tubule. *Am. J. Physiol.* 233: F359–65.
Satchell, G. H. 1961. The response of the dogfish to anoxia. *J. Exp. Biol.* 38: 531–43.
– 1976. The circulatory system of air-breathing fish. In *Respiration of Amphibious Vertebrates,* ed. G. H. Hughes, pp. 105–24. London: Academic Press.
Scheid, P. and J. Piiper. 1976. Quantitative functional analysis of branchial gas transfer: Theory and application to *Scyliorhinus stellaris* (Elasmobranchii). In *Respiration of Amphibious Vertebrates,* ed. G. M. Hughes, pp. 17–38. London: Academic Press.
Schidlowski, M. 1975. Archaean atmosphere and evolution of the terrestrial O_2 budget. In *The Early History of the Earth,* ed. B. F. Windley. London: John Wiley and Sons.
Schwartz, J. H. 1976. H^+ current response to CO_2 and carbonic anhydrase inhibition in turtle bladder. *Am. J. Physiol.* 231: 565–72.
Shelton, G. 1970a. The effect of lung ventilation on blood flow to the lungs and body of the amphibian, *Xenopus laevis*. *Respir. Physiol.* 9: 183–96.
– 1970b. The regulation of breathing. In *Fish Physiology,* ed. W. S. Hoar and D. J. Randall, vol. 4, pp. 293–359. New York: Academic Press.
Shelton, G. 1975. Gas exchange, pulmonary blood supply, and the partially divided amphibian heart. In *Perspectives in Experimental Biology*, ed. P. Spencer Davies, vol. 1, pp. 247–59. Oxford: Pergamon Press.
Shelton, G. and W. Burggren. 1976. Cardiovascular dynamics of the chelonia during apnoea and lung ventilation. *J. Exp. Biol.* 64: 323–43.
Singh, B. N. and G. M. Hughes. 1971. Respiration of an air-breathing catfish, *Clarias batrachus* (Linn.). *J. Exp. Biol.* 55: 421–34.
– 1973. Cardiac and respiratory responses in the climbing perch *Anabas testudineus*. *J. Comp. Physiol.* 84: 205–26.
Smith, D. G. 1976. The innervation of the cutaneous artery in the toad, *Bufo marinus*. *Gen. Pharmacol.* 7: 405–9.
Smith, D. G. and B. J. Gannon. 1978. Selective control of branchial arch perfusion in an air-breathing Amazonian fish *Hoplerythrinus unitaeniatus*. *Can. J. Zool.* 56: 959–64.
Smith, D. G. and D. H. Macintyre. 1979. Autonomic innervation of the visceral and vascular smooth muscle of a snake lung (Ophidia: Colubridae). *Comp. Biochem. Physiol.* 62C: 187–91.
Smith, F. M. and D. R. Jones. 1978. Localization of receptors causing hypoxic bradycardia in trout (*Salmo gairdneri*). *Can. J. Zool.* 56: 1260–5.
Smith, H. W. 1929. The excretion of ammonia and urea by the gills of fish. *J. Biol. Chem.* 81: 727–42.
Sobin, S. S., Y. C. Fung, H. M. Tremer, and T. H. Rosenquist. 1972. Elasticity of the pulmonary alveolar microvascular sheet in the cat. *Circ. Res.* 30: 440–50.
Stevens, E. D. and G. F. Holeton. 1978. The partitioning of oxygen uptake from

air and from water by the large obligate air-breathing teleost, pirarucu (*Arapaima gigas*). *Can. J. Zool.* 56: 974–6.

Tappan, H. 1974. Molecular oxygen and evolution. In *Molecular Oxygen in Biology*, ed. O. Hayaishi, pp. 81–135. Amsterdam, Oxford, and New York: North Holland/American Elsevier.

Thomas, K. S. 1969. The environment and distribution of Paleozoic sarcopterygian fishes. *Am. J. Sci.* 267: 457–64.

Todd, E. S. and A. W. Ebeling. 1966. Aerial respiration in the long jaw mudsucker *Gillichthys mirabilis* (Teleostei: Gobiidae). *Biol. Bull.* 130: 256–88.

Toews, D. P., G. Shelton, and D. J. Randall. 1971. Gas tensions in the lungs and major blood vessels of the urodele amphibian, *Amphiuma tridactylum*. *J. Exp. Biol.* 55: 47–61.

Webb, G., H. Heatwole, and J. De Bavay. 1971. Comparative cardiac anatomy of the Reptilia. I. The chambers and septa of the varanid ventricle. *J. Morphol.* 134: 335–50.

West, J. B. 1974. *Respiratory Physiology – The Essentials*. Baltimore: The Williams & Wilkins Co. 185 pp.

– 1977. *Ventilation/Blood Flow and Gas Exchange*, 3rd ed. Oxford: Blackwell Scientific Publ. 113 pp.

West, N. H. and D. R. Jones. 1975a. Breathing movements in the frog, *Rana pipiens*. I. The mechanical events associated with lung and buccal ventilation. *Can. J. Zool.* 53: 332–44.

– 1975b. Breathing movements in the frog, *Rana pipiens*. II. The power output and efficiency of breathing. *Can. J. Zool.* 53: 345–53.

White, F. N. 1976. Circulation. In *Biology of the Reptilia*, ed. C. Gans, vol. 5, pp. 275–334. New York: Academic Press.

White, F. N. and G. Ross. 1966. Circulatory changes during experimental diving in the turtle. *Am. J. Physiol.* 211: 15–18.

Willmer, E. N. 1934. Some observations on the respiration of certain tropical fresh water fish. *J. Exp. Biol.* 11: 283–306.

Wood, C. M. and F. H. Caldwell. 1978. Renal regulation of acid–base balance in a freshwater fish. *J. Exp. Zool.* 205: 301–7.

Wood, C. M. and D. J. Randall. 1973. The influence of swimming activity on water balance in the rainbow trout (*Salmo gairdneri*). *J. Comp. Physiol.* 82: 257–76.

Index